W9-DFR-709

MATHEMATICS
EXPLAINED

MATHEMATICS EXPLAINED

MARK FRARY

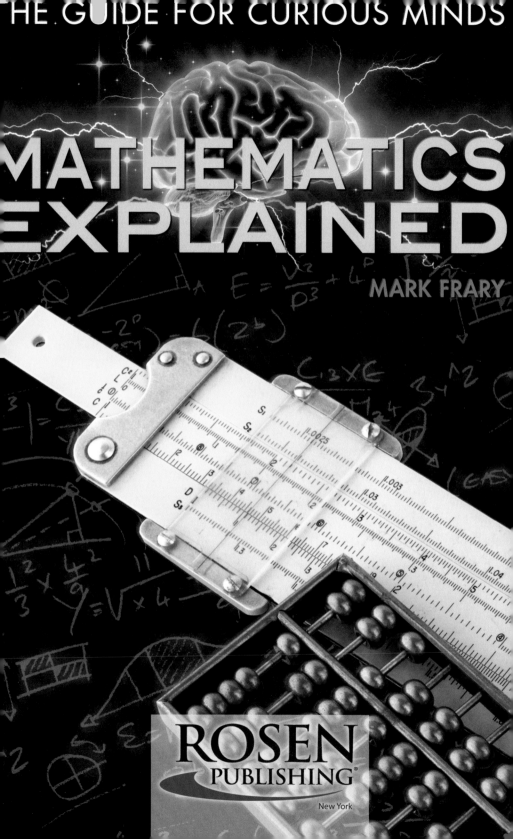

ROSEN
PUBLISHING®

New York

This edition published in 2014 by:

The Rosen Publishing Group, Inc.
29 East 21st Street
New York, NY 10010

Additional end matter copyright © 2014 by The Rosen Publishing Group, Inc.

Library of Congress Cataloging-in-Publication Data

Frary, Mark.
Mathematics explained/Mark Frary.—First edition.
    pages cm.—(The guide for curious minds)
Audience: Grade 7 to 12.
Includes bibliographical references and index.
ISBN 978-1-4777-2971-7 (library binding)
1. Mathematics—Miscellanea—Juvenile literature. I. Title.
QA40.5.F73 2014
510—dc23

                                                    2013030557

Manufactured in the United States of America.

CPSIA Compliance Information: Batch #W14YA: For further information, contact Rosen Publishing, New York, New York, at 1-800-237-9932.

© 2014 ELWIN STREET PRODUCTIONS
Elwin Street Productions
3 Percy Street
London W1, UK
www.elwinstreet.com

Contents

Puzzlebox solutions

These appear between pages 122 through 124.

THE STORY OF
NUMBERS

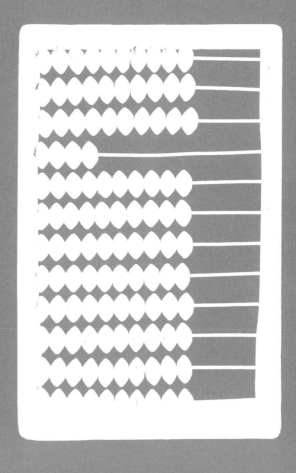

What are numbers?

It is hard to imagine a world without numbers. The evidence suggests that numbers were in use by 3000 BC and possibly much earlier. At that time, numbers were used to indicate quantities of specific objects – three oxen, for example. But there were no specific symbols to represent numbers. Instead, people used repeated marks, called tally marks, to represent quantity – four marks to represent four, and so on.

The first numerals

The earliest evidence of numerals can be found in carvings in what is now eastern Afghanistan and the northern part of the Punjab. Kharosthi was an alphanumerical system used in the region until at least the third century AD. Over time, the representation of numbers became more sophisticated. Many cultures used letter characters or just spelled out the words for one, two, three, etc. But this makes doing anything with them – such as math – virtually impossible.

The source of modern numbers

Many believe that the numerical symbols used today originate from Hindu mathematicians in India around the third century BC (see chart, page 9).

By the first century BC, these symbols had developed, as the Saka numerals in the table show. The Saka were a tribe of nomads living in what is now Iran.

By 200 AD, the relationship to our current-day numbers becomes even clearer, as demonstrated by the numerals found in a cave system in in Maharashtra, India.

By the 10th century AD, the numbers are clearly similar to our own. The gobar numerals are thought to have been brought to Spain by Arab mathematicians who had learned of them from Indian mathematicians.

Dated to around 20,000 BC, the Ishango bone was found in Africa and may be the world's oldest mathematical artifact. It shows a series of markings that appear to represent numbers.

Numbers through the ages

Kharosthi numerals 300 BC	Ashoka numerals 300 BC	Saka numerals 100 BC	Nasik numerals 200 AD	Roman numerals 100 BC/100 AD –present	Gobar numerals 900 AD	Eastern Arabic numbers c.900 AD –present	Hindu-Arabic numbers c.900 AD –present
						•	0
I	I	I	—	I	I	ﻝ	1
II	II	II	=	II	٢	٢	2
		III	≡	III	٢	٣	3
IIII	+	X	⋎ ⵕ	IV or IIII	ⴌ	٤	4
IIIII		IX	ⴔⴄ	V	٩	٥	5
	ᴪᴪ	IIX	ⴖ	VI	٦	٦	6
			٩	VII	٢	٧	7
		XX	٤ⴘ	VIII	8	٨	8
			?	IX	٩	٩	9

Notes: Blank spaces indicate no known numeral. Note that most early systems did not have a zero (see later in this chapter). Some double characters show alternative representations. Source: *The Hindu-Arabic Numerals* by David Eugene Smith and Louis Charles Karpinski (1911).

FACT

Until the invention of the printing press, the only way to record and use symbols was by rewriting them, and small changes inevitably occurred during transcription.

Major mathematical centers

Babylon

This ancient city state between the Tigris and Euphrates in modern-day Iraq, which existed from around 3000 BC to 540 BC, was a hotbed of mathematical activity. The Babylonians came up with a

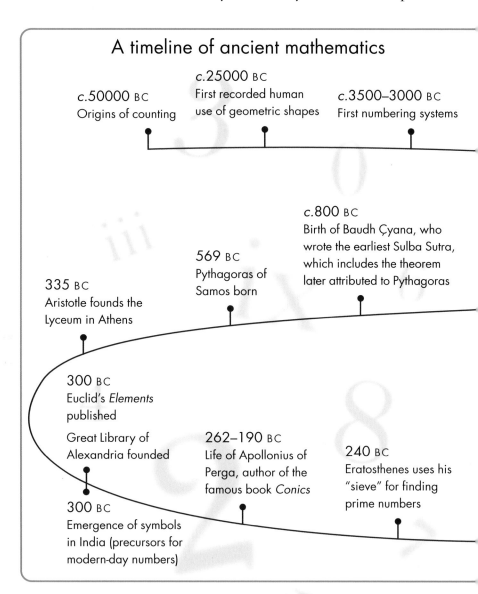

A timeline of ancient mathematics

c.50000 BC
Origins of counting

c.25000 BC
First recorded human use of geometric shapes

c.3500–3000 BC
First numbering systems

335 BC
Aristotle founds the Lyceum in Athens

569 BC
Pythagoras of Samos born

c.800 BC
Birth of Baudh Çyana, who wrote the earliest Sulba Sutra, which includes the theorem later attributed to Pythagoras

300 BC
Euclid's *Elements* published

Great Library of Alexandria founded

300 BC
Emergence of symbols in India (precursors for modern-day numbers)

262–190 BC
Life of Apollonius of Perga, author of the famous book *Conics*

240 BC
Eratosthenes uses his "sieve" for finding prime numbers

numerical system based on the number 60, a so-called sexagesimal system (as opposed to the 10-based decimal system we use today).

The region's mathematicians also worked with fractions and had some knowledge of solving simple quadratic equations. A clay tablet dating from around 1800 BC and known as Plimpton 322 (the 322nd in the G.A. Plimpton Collection at Columbia University) appears to show a list of numbers that solve the equation $x - 1/x = c$.

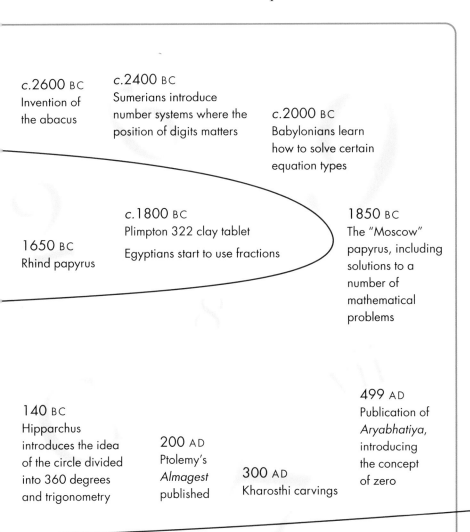

c.2600 BC
Invention of the abacus

c.2400 BC
Sumerians introduce number systems where the position of digits matters

c.2000 BC
Babylonians learn how to solve certain equation types

c.1800 BC
Plimpton 322 clay tablet

Egyptians start to use fractions

1850 BC
The "Moscow" papyrus, including solutions to a number of mathematical problems

1650 BC
Rhind papyrus

140 BC
Hipparchus introduces the idea of the circle divided into 360 degrees and trigonometry

200 AD
Ptolemy's Almagest published

300 AD
Kharosthi carvings

499 AD
Publication of Aryabhatiya, introducing the concept of zero

Egypt

At around the same time that Plimpton 322 was being made, the Egyptians were writing down their own mathematical ideas on papyrus. They, too, were using fractions, but only unit fractions (i.e. where the numerator of the fraction is 1, such as ½ and ¼).

The Rhind papyrus—named after Egyptologist Alexander Rhind, who bought it in the mid-19th century—dates from around 1650 BC and shows that the Egyptians had extensive mathematical knowledge, including an understanding of geometry and algebra.

As with the Babylonians, the Egyptians seem to have used mathematics for practical problems, such as working out how to share out food or calculating the area and volume of certain objects.

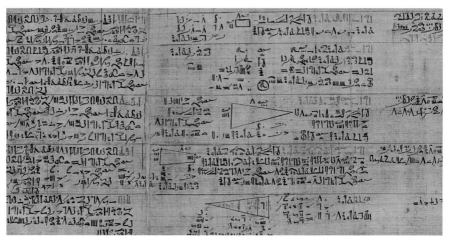

Different parts of the Rhind papyrus (above) dealt with areas of rectangles, triangles, and pyramids.

Ancient Greece

The ancient Greeks, by contrast, appear to be the first people who studied mathematics for its own sake, rather than to solve problems. Their principal achievements were in the areas of geometry and number theory, which looks at the special properties of numbers.

The Greeks were also the first to look at the concept of mathematical proof, the idea of being able to prove through deduction that a mathematical statement is true for every case.

FACT

From around the sixth century BC to the fifth century AD, ancient Greece was home to some of the most important mathematicians of the era. They included Pythagoras, famous for his theorem of triangles, Thales and Euclid, for their study of geometry, and Archimedes (of eureka fame), for his work with mathematical limits.

China

Much of China's early mathematical history is tied up with the mechanization of counting. Counting rods – small bars that represented numbers – were in use around 2,000 years ago in China and the rest of Asia.

The Nine Chapters on the Mathematical Art was continually developed by Chinese mathematicians between 1000 BC and 100 AD and shows, among other things, that the Chinese understood how to calculate areas, volumes, and proportions, and had a working knowledge of algebra.

India

Indian mathematics really began, as far as we can tell, in the Indus Valley between 2500 BC and 1700 BC. In this area, people were using decimal fractions to represent weights and lengths.

A thousand years later, followers of the Vedic religion were transmitting their knowledge of the construction of altars (including what we know as Pythagoras's theorem) through oral recitation of the Sulba Sutras.

In 1881, a mathematical work now known as the Bakhshali Manuscript was discovered in India. Believed to date from the second or third century AD and written on pieces of birch bark, it included discussions of square roots and the use of negative numbers.

Great mathematicians through the ages

Brahmagupta
(598–668) India
Zero, use of negative numbers, linear equations

Muhammad ibn Musa Khwarizmi
(780–850) Uzbekistan
Algebra (the word comes from one of his books)

Al-Karaji
(953–1029) Persia
Proof by induction, algebra

Omar Khayyam
(1048–1123) Persia
Algebra (including the use of x to represent an unknown quantity), theory of parallels

Adelard of Bath
(1080–1152) England
Translation of key Arabic works into Latin

Bhaskara
(1114–1185) India
Quadratic and cubic equations, differential calculus, division by zero, use of decimal system

Sharaf al-Din al-Tusi
(1135–1213) Persia
Concept of the function, cubic equations

Fibonacci
(1170–1250) Italy
Fibonacci numbers and popularizing Hindu-Arabic numbering system

Nicole Oresme
(1323–1382) France
Coordinate geometry, fractional powers

Scipione del Ferro
(1465–1526) Italy
Solution of cubic equations

Robert Recorde
(1510–1558) Wales
Introduced the equals sign

Lodovico Ferrari
(1522–1565) Italy
Quartic equations

Bartholomaeus Pitiscus
(1561–1613) Germany
Trigonometry

William Oughtred
(1575–1660) England
Introduced multiplication symbol and the use of sin and cos

René Descartes
(1596–1650) France
Analytic geometry, study of tangents, laid groundwork for calculus

Pierre de Fermat
(1601–1665) France
Theory of numbers, laid groundwork for calculus, known for Fermat's last theorem (no three positive integers a, b, and c can satisfy $a^n + b^n = c^n$ for n greater than two)

John Wallis
(1616–1703) Great Britain
Powers notation, number lines, infinite series

Blaise Pascal
(1623–1662) France
Pascal's triangle of binomial coefficients, probability theory

Johan De Witt
(1625–1672) Netherlands
Probability theory

Isaac Newton
(1642–1727) Great Britain
Infinitesimal calculus, binomial theorem, power series

Gottfried Wilhelm Leibniz
(1646–1716) Germany
Infinitesimal calculus, topology, linear equations

Leonhard Euler
(1707–1783) Switzerland
Mathematical notation (e, i, functions), power series, number theory

Maria Agnesi
(1718–1799) Italy
Differential and integral calculus

Joseph-Louis Lagrange
(1736–1813) Italy
Calculus of variations, Lagrangian mechanics, probability, group theory

Note: Some dates are approximate.

Great mathematicians of the period 1650–1900

Pierre Simon de Laplace
(1749–1827) France
Statistical and probability theory, differential equations

Adrien-Marie Legendre
(1752–1833) France
Least squares method, statistics, elliptic functions, Legendre transformation

Carl Friedrich Gauss
(1777–1855) Germany
Number theory, statistics, fundamental theorem of algebra

Augustin-Louis Cauchy
(1789–1857) France
Infinitesimal calculus, analysis, complex function theory

August Möbius
(1790–1868) Germany
Möbius strip, number theory

Charles Babbage
(1791–1871) Great Britain
Designed difference and analytical engines (forerunners of modern computers)

Carl Jacobi
(1804–1851) Prussia
Elliptic functions, number theory

Augustus De Morgan
(1806–1871) Great Britain (born India)
Laws of algebra, mathematical induction

Evariste Galois
(1811–1832) France
Group theory

George Boole
(1815–1864) UK
Logic

Francis Galton
(1822–1911) UK
Standard deviation, regression

Bernhard Riemann
(1826–1866) Germany
N-dimensional geometry, analysis

Sofya Kovalevskaya
(1850–1891) Russia
Analysis, differential equations

Henri Poincaré
(1854–1912) France
Poincaré conjecture, mathematical physics

Andrei Markov
(1856–1922) Russia
Stochastic processes

David Hilbert
(1862–1943) Germany
Invariant theory, functional analysis, Hilbert spaces

Amalie Emmy Noether
(1882–1935) Germany
Abstract algebra

Robert L. Moore
(1882–1974) USA
Topology

Hermann Weyl
(1885–1955) Germany
Manifolds and topology

Norbert Wiener
(1894–1964) USA
Set theory

Mary Cartwright
(1900–1998) UK
Function theory

John von Neumann
(1903–1957) Hungary
Logic, set theory and game theory

Andrei Kolmogorov
(1903–1987) Russia
Topology, probability theory, logic

Kurt Gödel
(1906–1978) Austro-Hungarian Empire
Incompleteness theorems, mathematical proof

Alan Turing
(1912–1954) UK
Algorithms, computability

Paul Erdös
(1913–1996) Austro-Hungarian Empire
Number theory, combinatorics, probability

John Tukey
(1915–2000) USA
Statistics

Julia Robinson
(1919–1985) USA
Decision problems

Benoît Mandelbrot
(1924–) France
Fractal geometry

Andrew John Wiles
(1953–) UK
Proving Fermat's last theorem, number theory

Counting machines

Mathematicians and mere mortals have used a variety of devices to help them with their numbers throughout history.

The abacus

The abacus, whose name comes from the ancient Greek *abax*, or "board," is still seen today. We do not know precisely who invented the abacus but many point to the Sumerians, who lived in what is now southern Iraq. Its design has remained largely unchanged: a frame holding columns of beads, with each column representing a certain multiple of the previous. For the Sumerians and their sexagesimal numbering system, this multiple was 60; for those using decimals, the multiple is 10.

The abacus was the first true calculating machine, and was probably invented by the Sumerians sometime around 2600 BC.

The Antikythera mechanism

The abacus is a simple device, which is not something that can be said of the Antikythera mechanism. Made of 30 intricate interlocking gears, this unusual device was found on a shipwreck off the island of Antikythera in Greece in 1891. In recent years, researchers have postulated that it is a highly accurate computer, used for making astronomical predictions. It was probably made between 150 BC and 100 BC and shows that the Greeks had much more sophisticated understanding of mathematics than previously thought.

The calculating clock

Much later on, in 1623, the German Wilhelm Schickard invented the forerunner of the calculator. The calculating clock, which was constructed using cogs intended for use in clocks, could add and subtract six-digit numbers.

Pascal's machine

French scientist and philosopher Blaise Pascal produced this mechanical adding machine between 1642 and 1645. Similar to the Schickard device in construction, it could perform addition and subtraction over eight columns of digits.

An example of Pascal's calculating machine, produced in the 17th century.

Leibniz's calculating machine

Gottfried Leibniz (a German mathematician noted for his development of calculus) came up with a considerably more sophisticated calculator than Pascal's in 1673. It could multiply, divide, and find square roots.

The difference engine

In 1822, British mathematician Charles Babbage suggested building a steam-powered mechanical computer made of brass to produce logarithm tables automatically and thus eliminate all sources of inaccuracy. Despite receiving substantial funding from the British government, Babbage never built his so-called difference engine but instead went on to make plans for a more general-purpose mechanical computer, the analytical engine. A working difference engine was built, using Babbage's designs, by London's Science Museum in 1991.

The first pocket calculators

The first truly pocket-sized calculator, the Busicom LE-120A, appeared in 1971. It had a 12-digit display and could only work with a fixed decimal point. The following year, Hewlett Packard launched the first scientific pocket calculator, the HP35, which could calculate sines and cosines. Until then, calculators could only add, subtract, divide, and multiply. On initial release, both calculators cost $395.

Engineers at London's Science Museum constructed their difference engine from designs made by Charles Babbage between 1847 and 1849. He conceived the engine to calculate a series of numerical values and automatically print the results.

MATHEMATICAL
CONCEPTS

Mathematical symbols

The use of a number system based on the number 10 seems to be so logical that it is hard to imagine using anything else. It is often thought that the decimal system developed because we have 10 fingers and thumbs but that doesn't explain why the great mathematicians of ancient Babylonia used a sexagesimal numbering system based on 60.

Zero

Although their numbering system has not continued into modern practice, another Babylonian mathematical concept has – the zero. Why do we need it? Perhaps the most powerful aspect of the zero is that you do not need to invent new symbols or conventions to represent ever larger numbers, as the Romans did.

$$I = 1$$
$$X = 10$$
$$C = 100$$
$$M = 1000$$

The problems with the Romans' system come when you try to make calculations. It is not easy to add and subtract Roman numerals on paper.

Zero helps overcome these limitations because it can be used as a placeholder, freeing the digits 1 to 9 to be used as an indicator of quantity. Thus the digit 3 can be used to represent 3, 30, 300, 3,000 and, indeed, a limitless set of numbers differentiated only by an increasing number of zeros. Zero as a placeholder also makes handling numbers, however large, considerably easier.

Negative numbers

Negative numbers have been around since the fourth century BC and probably date from an even earlier time. We know this because Chinese counting rods were found in a grave dating from the

FACT
The concept of zero as a number in its own right was not considered until Brahmagupta established arithmetical rules for using it in the seventh century AD. He set out rules that adding zero to zero gave zero and multiplying any number by zero also gave zero.

Warring States Period in the country's history. These small rods are used for calculation and are colored red for positive numbers and black for negative numbers.

Brahmagupta's rules
The Indian mathematician Brahmagupta set out formal rules on how to handle negative numbers (which he called debts) and positive numbers (fortunes).

Brahmagupta's rule	What it means
A debt subtracted from zero is fortune	Subtracting a negative number from zero gives a positive number
A fortune subtracted from zero is a debt	A positive number subtracted from zero gives a negative number
The product or quotient of two debts is a fortune	Two negative numbers multiplied gives a positive number
The product or quotient of a debt and a fortune is a debt	A negative number multiplied by or divided by a positive number gives a negative number

Natural numbers
Natural numbers is the name mathematicians give to the everyday numbers we use for counting and ordering, e.g. Houston is the fourth largest city by population in the U.S. Some mathematicians include

zero among the natural numbers, although others do not, which has led to a heated (and as yet unresolved) debate.

Integers

Integers are whole numbers (i.e. not those involving fractions or decimals). This means they encompass all of the natural numbers and their negative counterparts, e.g. -4, -3, -2, -1, 0, 1, 2, 3, 4, and so on. Adding, subtracting, and multiplying any two integers will always give another integer.

Basic arithmetical concepts

Equations and formulas

An equation is a mathematical statement where one side is equal to the other. A formula is a broader concept, which can include things other than the idea of equality. Equations and formulas commonly involve not just numbers but also letters that stand in the place of numbers, for example $y = x + 2$, or $y > x + 2$.

Multiplication and addition

It's easy to see that $1 + 2$ gives the same answer as $2 + 1$. The same is true of any two chosen numbers. If we say a and b are the two numbers, then we can say that $a + b = b + a$, no matter what values of a and b you choose. This is known as the law of commutation.

Multiplication is also commutative, since $a \times b = b \times a$. But the same cannot be said of subtraction ($a - b \neq b - a$) nor of division ($a \div b \neq b \div a$) unless a is the same number as b.

Multiplication and addition also obey the law of association: a + $(b + c) = (a + b) + c$ and a $(b \times c) = (a \times b)$ c. This means that no matter which order you work out the sum, you get the same answer. Again, this does not apply to subtraction and division.

These two laws are extremely helpful when it comes to algebra: they allow us to reorganize equations into more easily handled ones.

Division

Each term of a division sum has a specific name. The number being divided is called the dividend while the number used to divide by is the divisor. The result is called the quotient. If the divisor doesn't fit into the dividend a whole number of times, the amount left over is called the remainder.

The remainder can be written out expressly (for example, $5 \div 2 = 2$ remainder 1). For exact answers, it can be converted into a fraction of the divisor, which can also be expressed as a decimal.

Squares and indices

A power or index (plural: indices) can be used as a shorthand for representing large numbers and complex sums. The first experience many people have of indices is through squares. When you square a number, you multiply it by itself, e.g. 3 squared is 3×3. Mathematicians also write this as 3^2 where the small 2 shows how many times you multiply the 3 together. That small 2 is what we call an index or power.

Cubes

Cubing a number means multiplying it together three times; 4 cubed is $4 \times 4 \times 4$ or 4^3. Here the small 3 is the index or power while the big number 4 is what is known as the base. The commonly used term "y to the power x" is a shorthand way of saying y multiplied by itself x times.

Powers of 10

Powers of 10 can be used to express very large numbers. The number 100 is the same as 10 squared or 10^2. Similarly 1,000 is 10^3, 10,000 is 10^4, 100,000 is 10^5, and 1,000,000 is 10^6. As numbers get longer, it gets easier to use powers of 10. Writing 10^{21} is much quicker than 1,000,000,000,000,000,000,000.

A form of this notation using negative numbers can also be used for really small fractions. A half, ½, can also be referred to as 2 to the power of –1 or 2^{-1}. Similarly a quarter can be thought of as 1 divided by two squared or 2^{-2}.

The same goes for powers of 10. 10^{-2} means one divided by 10 squared. 10^{-3} means one divided by 10 cubed. Any very small number can be written using these powers of 10. 1.3×10^{-21} is the same as writing 1.3 times one divided by 1,000,000,000,000,000,000,000 or 0.0000000000000000000013.

Superpowers

Powers of 10 can be used to represent
any number, not just round billions, trillions, and gazillions:
You can multiply them by a number between 1 and 10 to get
other numbers. For example 1.3×10^{21} is the same as
1,300,000,000,000,000,000,000.

How would you write 7,354,267 using powers of 10?

Roots

Roots are the converse of powers. If 3 squared is the same as 3^2 or 3×3, or 9, then looking at this the other way round, 3 is the square root of 9 or $\sqrt{9}$. Similarly, if 3^3 is the same as $3 \times 3 \times 3$, or 27, it is also true that 3 is the cube root of 27.

Powers can also be used to represent roots. A square root of a number is the same as raising that number to the power of ½ while a cube root is the same as raising the number to the power of ⅓.

Although you usually see only square and cube roots, it is possible to see and use fourth, fifth, and sixth roots of numbers. It is also mathematically possible to use a one millionth root of a number.

Powers of 10 number line

10^{-15}	Size of a proton
10^{-14}	Range of the weak nuclear force
10^{-13}	Duration of a fast chemical reaction
10^{-12}	Weight of a human body cell
10^{-11}	Time after the Big Bang when electromagnetism separates from other fundamental forces
10^{-10}	Rest mass energy of a neutron
10^{-9}	Size of a typical virus
10^{-8}	Weight of a grain of sand
10^{-7}	Wavelength of visible light
10^{-6}	Volume of liquid held by a teaspoon
10^{-5}	Diameter of a blood cell
10^{-4}	Width of a human hair
10^{-3}	Diameter of a mustard seed
10^{-2}	Weight of a small bird
10^{-1}	Diameter of a tennis ball
10^{0}	Height of a human
10^{1}	The world record for the 100 m
10^{2}	Length of a football pitch
10^{3}	Weight of a typical small family car
10^{4}	Depth of the deepest ocean on Earth
10^{5}	Length of a day
10^{6}	Weight of the Space Shuttle at launch
10^{7}	Distance around the equator
10^{8}	Half-life of cobalt-60, one of the by-products of nuclear weapons
10^{9}	Weight of the Great Pyramid of Giza
10^{10}	The distance light travels in a minute
10^{11}	Distance from Earth to the Sun
10^{12}	Estimate of number of fish in the world's oceans
10^{13}	Distance travelled by Voyage spacecraft
10^{14}	Yield of the Fat Man bomb at Nagasaki
10^{15}	Surface area of Neptune
10^{16}	Distance to Proxima Centauri, the nearest star after the Sun
10^{17}	Age of the universe

10^{9} Weight of the Great Pyramid of Giza

10^{3} Weight of a typical small family car

10^{6} Weight of the Space Shuttle at launch

10^{-2} Weight of a small bird

10^{-12} Weight of a human body cell

Approximate orders of magnitude shown, lengths given in meters, areas in square meters, volumes in cubic meters, weights in kilograms, times in seconds, energies in joules.

Fractions

Fractions comprise two parts: the numerator – the number on top of the fraction – and the denominator – the number on the bottom.

A useful skill to master is to be able to simplify complex fractions. Consider the fraction ¾ or two quarters. If we divide a pie up into quarters and color in two of them, we get the picture below.

Clearly, two quarters is just the same as a half. The fraction can be simplified by using some basic rules. The denominator, 4, is the same as 2 × 2 and the numerator, 2, the same as 2 × 1, so we can rewrite it as follows:

$$\frac{2}{4} = \frac{2 \times 1}{2 \times 2}$$

We can now do something called cancellation. There are 2s on both the top and the bottom of the right-hand side of this equation. These two cancel each other out, leaving the simplified fraction ½. This cancellation method works for any fraction that can be rewritten to have the same digits at the top and the bottom.

Reciprocal division

The reciprocal method is helpful when you are working with fractions. The rule is: Multiply the dividend by the reciprocal of the divisor. (The reciprocal of a number is 1 divided by that number.)

What is 12 divided by ⅓?

Decimals

Fractions are useful for certain types of numbers between zero and one but cannot be easily used for more complicated numbers. The decimal system is based on the number 10 and its powers, and can also represent numbers smaller than 1. If you divide 12,345 by 10, you get the answer 1,234 remainder 5, or 1,234 plus $\frac{5}{10}$. To convert this to a decimal, we put a decimal point down after 1,234 and then write down how many tenths there are. Thus, the answer is 1,234.5.

Small numbers can be similarly represented. In the same way that each position we move to the left of the decimal point gets bigger by a power of 10, each position to the right moves down a power of 10. This means that the first position after the point is the number of tenths, the next one is the number of hundredths and so on.

So 8.7654 means 8 ones plus 7 tenths plus 6 hundredths plus 5 thousandths plus 4 ten thousandths.

Decimal places

Rounding (the process of approximating a number to the nearest easy number such as 10, 100, or 1,000,000) can also be done with digits after the decimal place: in effect, rounding to the nearest tenth, hundredth, or millionth. This is expressed as giving the number rounded to 1, 2, or 6 decimal places, or however many digits are given after the decimal point.

Significant figures

Another rounding technique, this is useful for calculating the accuracy and precision of measurements. Consider the sum 123 divided by 100,000, which gives 0.00123. The zeros immediately after the decimal point do not tell us anything about the number other than how small it is. They are said to be insignificant. Instead of quoting to a number of decimal places, we can quote it to a certain number of significant figures. 0.00123 to one significant figure is 0.001, to two significant figures it is 0.0012, and so on.

Percentages

It can be useful to calculate how much something has increased or decreased compared to a base value. For example, say a child is 24 inches (60 cm) tall on his second birthday and 28 inches (72 cm) on his third. We can easily work out that he has grown 4 inches (12 cm), but that does not tell us much. We need to make a comparison to the original height.

One way to do this is to calculate a percentage. Divide the original amount by 100 and call this 1 percent (or 1%) of the original amount. Rather than adding these 1%s by trial and error, the following equation can be used:

$$\frac{\text{Percentage change}}{\text{in a quantity}} = \frac{\text{New quantity} - \text{Old quantity}}{\text{Old quantity}} \times 100\%$$

$$\frac{\textbf{PERCENTAGE CHANGE}}{\text{in height}} = \frac{(28 - 24) \times 100\%}{24} = 16.6\%$$

(Note: A common mistake when using this equation is to put the new, rather than the old, quantity in the denominator of the fraction.)

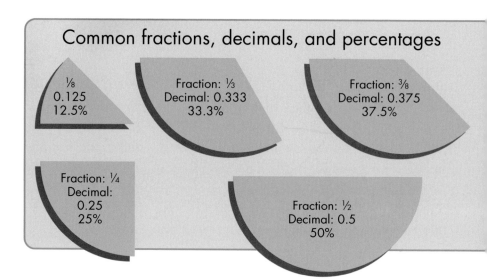

Common fractions, decimals, and percentages

⅛
0.125
12.5%

Fraction: ⅓
Decimal: 0.333
33.3%

Fraction: ⅜
Decimal: 0.375
37.5%

Fraction: ¼
Decimal:
0.25
25%

Fraction: ½
Decimal: 0.5
50%

Rational and irrational numbers

Irrational numbers are those that cannot be written as simple fractions such as *a/b*, where *a* and *b* are whole numbers, nor as decimals with recurring or repeating digits. The square root of two and the special numbers pi and *e* are examples of irrational numbers. All other numbers that can be represented as a simple fraction or a non-recurring decimal are called rational numbers.

Special numbers

Pi, π

Pi is perhaps the best known of all the special numbers. It is defined as the circumference of a circle (the length of its edge) divided by its diameter (the distance from one edge to the other passing through the center). It has a value of 3.1415926535… The ellipsis here shows that pi is irrational (see above) and the part beyond the decimal point continues indefinitely, without repetition. An approximate figure of 3 was used in calculations until the third century BC, when Archimedes established a figure of 3.14 for pi. This was then refined to 3.1416 during the second century AD, and gradually moved to a greater number of decimal places as the centuries passed.

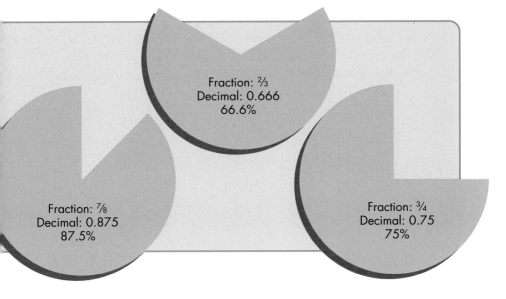

Fraction: ⅔
Decimal: 0.666
66.6%

Fraction: ⅞
Decimal: 0.875
87.5%

Fraction: ¾
Decimal: 0.75
75%

Infinity

Infinity is not strictly a number, but it is useful to mention it here. It is represented by the symbol ∞ and is used to show the concept of an unbounded limit. For example, you could say that infinity is the sum of this never-ending series:

$$1 + 2 + 4 + 8 + 16 + 32 + \dots$$

It is often used in the understanding of limits. Consider the reciprocal function $1/x$. As x gets larger and more positive, $1/x$ gets smaller and smaller. A graph of $1/x$ shows it increasingly close to zero as x gets larger and larger. Mathematicians write this as:

$$\lim_{x \to \infty} 1/x = 0$$

and say that the limit of $1/x$ as x increases to infinity is zero.

Prime numbers

Prime numbers are numbers that can only be divided (without leaving a remainder) by themselves and the number one. The first few primes are as follows:

$$1, 2, 3, 5, 7, 11, 13, 17, 19, 23 \dots$$

The Greek mathematician Euclid proved that there are an infinite number of prime numbers while another Greek mathematician, Eratosthenes, worked out a method for finding prime numbers. It was called the Sieve of Eratosthenes and involves writing down a list of numbers and then crossing out multiples of two, three, and so on all the way up to the square root of the highest number in the list.

In addition to his work on prime numbers, Erastothenes was the first person to calculate Earth's circumference.

FACT

The biggest prime number found at the date of publication has a whopping 7 million digits and was discovered using a piece of software called GIMPS, which anyone with a modern computer can download.

Perfect numbers

What defines perfection? For a mathematician, a perfect number has a well-defined mathematical property. It is any number whose divisors (other than the number itself) add together to make that same number.

Take the number 6, for example. Its divisors are 1, 2, 3, and 6. If we add together all of those other than the number 6, we get 1 + 2 +3 = 6. Hence, 6 is a perfect number.

The first four perfect numbers are 6, 28, 496, and 8,128. All of the perfect numbers found to date end in either a 6 or an 8. There is some debate over odd perfect numbers. No one has yet been able to prove conclusively that they don't exist, although they have checked all the way up to 10^{300}.

Euler's number

Another important number in mathematics is known as *e*, or Euler's number, after the Swiss mathematician Leonhard Euler. It is equal to 2.71828 to the first five decimal places, although it is irrational, meaning that the bit after the decimal point goes on forever. Mathematicians, with the help of computers, have calculated *e* to 100 billion decimal places. More exactly, it is equal to the following:

$$e = \lim_{n \to \infty} \left(1 + \frac{1}{n}\right)^n$$

The strange looking thing on the right hand side is known as a limit. Ignore this for the moment and just look at the thing in the brackets to the power *n*. Pick a value of n, say 2, and work it out. With *n=2*, you get:

$$(1 + \tfrac{1}{2})^2$$

which you can work out with your calculator is equal to 2.25
Now try it with *n* equal to 5

$$(1 + \tfrac{1}{5})^5$$

which you can work out is equal to 2.48832. With *n* equal to 20,000, the answer is 2.718214. Is that starting to look familiar? As *n* gets bigger, the closer the answer gets to the magic number *e*. If you were able to tap infinity into a calculator, the answer would come out as *e*.

FACT

The golden ratio (or mean) is an equation usually represented by the Greek letter phi or φ. It is approximately equal to 1.618. Some artists and architects believe that paintings and buildings (such as the Parthenon in Athens) proportioned according to the golden ratio are particularly pleasing.

BODMAS

What is the answer to the following sum? $8 + (5 \times 4^2 + 2)$

There is a standard convention among mathematicians, which is best known as BODMAS, a mnemonic standing for Brackets, Orders, Division, Multiplication, Addition, Subtraction. This mnemonic shows the order in which things should be calculated, i.e. you do the things in brackets first and the subtractions last. Orders are powers of numbers (see page 23).

Mathematical shortcuts

Faced with complicated multiplications, there are some quick shortcuts that will help you make an educated guess. This is particularly useful in multiple-choice situations.

What is $97 \times 1{,}014$?

a) 13,565   b) 56,018   c) 98,358

The first step is to look at the last digit of each number, i.e. 7 and 4. Multiplying those together gives you 28. The last digit of the answer to this easy sum is always the last digit of the more difficult sum. Knowing this means answer a) above can be discounted.

Best guess

The second step is to make a good guess of what the answer should be. The two numbers to be multiplied are 97 and 1,014. The first is very close to 100 and the second is very close to 1,000. Multiplying 100 and 1,000 is a very simple calculation to do in your head; it's 100,000. That means that the answer must be somewhere near 100,000. Answer c) is closest to this figure, so this would be the logical option to go for. Why not check it now on a calculator?

Averages

There are several different types of average value that mathematicians and statisticians use.

Mean

The mean is found by dividing the sum of quantities by the number of quantities, and is what people tend to mean when they say "average."

One of the problems with using the arithmetic mean is that it is prone to being skewed away from some central value. If you calculate the arithmetic mean of the salaries in a company, for example, the value can be skewed by a handful of directors earning vastly more than the typical employee.

Median

A median average is the central value in a list of numbers arranged in numerical order.

Mode

The mode of a sequence of numbers is the number that appears most frequently. For example, if we had a class of children and we wrote down all of their ages in a list as follows:

$$6, 6, 7, 7, 7, 7, 7, 7, 8, 8, 8$$

then the mode of this group of numbers is 7 because it appears more frequently than either 6 or 8.

Geometric mean

The geometric mean of a sequence of numbers is calculated by multiplying all of the numbers together and then taking the nth root.

$$\text{Geometric mean} = \sqrt[n]{a_1 \, a_2 \, a_3 \, \ldots \, a_n}$$

Fibonacci numbers

What is the next number in this sequence?

1   1   2   3   5   8   13   21   34   …

The answer is 55, but it is not obvious why. Examining the differences between the numbers within the sequence provides some insight:

Sequence

| 1 | | 1 | | 2 | | 3 | | 5 | | 8 | | 13 | | 21 | | 34 |…

Differences

| 0 | | 1 | | 1 | | 2 | | 3 | | 5 | | 8 | | 13 |…

The differences are the same (apart from an initial zero) as the original sequence. Thus we can guess that the next number should be 34 plus 21, or 55.

This sequence of numbers is known as the Fibonacci sequence, after 13th-century Italian mathematician Leonardo of Pisa (also known as Fibonacci), who introduced the idea to the West (it had been well known to Indian mathematicians long before). The series is used today in the financial markets, among other applications, and it also has surprising parallels in nature.

The Fibonacci series often tallies with the numbers of petals on flowers or segments in a fircone.

Number series

There is a type of sequence known as an arithmetic series or progression. It is a series of numbers where each number is larger (or smaller) than the previous number by a fixed amount.

There is a quick way of adding up any arithmetic series, which involves adding up the numbers in a different order. Addition follows the law of commutativity, i.e. $a + b = b + a$. This means that the sum can be written in any order and still have the same answer.

Add up the first number and the last number in the series. Now count how many numbers there are in the sequence. You then multiply those two numbers together and divide the result by two.

This works for any arithmetic series no matter how long and no matter what the difference is between successive terms.

Speedy series addition

How would you add up the following series of numbers quickly?

$$1 + 2 + 3 + 4 + 5 + 6 + 7 + 8 + 9 + 10 = ?$$

(Hint: Try writing out the sum, putting the first and the last numbers next to each other, the second and the second last, and so on. Then look at each consecutive pair of numbers.)

Infinitely long sums

If you could add up an infinitely long list of numbers, what would the answer be? You might think the answer would be infinity, but that's not always the case. Look at the following sum:

$$1 + 2 + 3 + 4 + 5 + 6 + 7 + 8 + \ldots$$

It's obvious here that the answer is infinite. A series that adds up to infinity is called a divergent series. However, there is another class of series that does not add up to infinity, but rather gets closer and closer to a specific value. These are called convergent series. The simplest example of a convergent series is perhaps the following:

$$1 + \tfrac{1}{2} + \tfrac{1}{4} + \tfrac{1}{8} + \tfrac{1}{16} + \tfrac{1}{32} + \tfrac{1}{64} + \ldots$$

In this series, each term is half the size of the previous one. So what does this sum equal? Mathematicians can prove that if you went on forever, the sum would eventually add up to 2.

An imaginative leap

Imagine jumping into a sandpit that is 6.5 feet (2 meters) across. On your first jump you jump 3.3 feet (1 meter) – halfway across. On your next jump from that midway point, you only manage to jump 1.6 feet (.5 meters). Each jump puts you slightly closer to the far side of the sandpit, but since each jump is always half of the remaining difference, you never quite reach it. If this sounds a little familiar it is because it is the principle behind what is known as Zeno's paradox (see page 38).

The importance of series

Certain infinite series converge on useful values, like that of pi or e:

$\pi/4 = \quad 1 - \tfrac{1}{3} + \tfrac{1}{5} - \tfrac{1}{7} + \tfrac{1}{9} - \tfrac{1}{11} + \ldots$

$\pi^2/12 = \quad 1 - \tfrac{1}{4} + \tfrac{1}{9} - \tfrac{1}{16} + \tfrac{1}{25} - \tfrac{1}{36} + \ldots$

$\sum \qquad\; 1 + \tfrac{1}{1!} + \tfrac{1}{2!} + \tfrac{1}{3!} + \tfrac{1}{4!} +$

$\sin x \qquad x - x^3/3! + x^5/5! - x^7/7! + \ldots$

$\cos x \qquad 1 - x^2/2! + x^4/4! - x^6/6! + \ldots$

The factorial sign ! means you should multiply all of the numbers from 1 to that number together, so $4! = 1 \times 2 \times 3 \times 4 = 24$. In the equations for sin and cos x, x is given in radians. Radians are a unit of angular measure, like the degree. There are 2π radians in a circle. Therefore, a radian is equal to about 57.3 degrees (since there are 360 degrees in a circle).

Zeno's paradox

Very little is known about the life of 5th-century BC philosopher Zeno of Elea, although his work is described by both Aristotle and Plato. Zeno's contribution to the world of mathematics was a series of paradoxes – a series of statements, which seem true but lead to an obviously false conclusion.

Achilles and the tortoise

This is Zeno's most famous paradox, as set out in Aristotle's *Physics*. In a race, the quickest runner can never overtake the slowest, since the pursuer must first reach the point from where the pursued started, so that the slower must always hold a lead.

Achilles is having a race with a tortoise. He runs at 16.4 feet (5 meters) per second and the tortoise at 1.64 feet (.5 meters) per second. Knowing that the tortoise is slower, he gives it a head start of 16.4 feet (5 meters). After 1 second, Achilles has run 16.4 feet (5 meters) and the tortoise has moved 1.64 feet (.5 meters). In the next tenth of a second, Achilles moves 1.64 feet (.5 meters), but the tortoise has already moved on 0.164 feet (.05 meters), so he has not caught it. In the next hundredth of a second, Achilles moves 0.164 feet (.05 meters), but the tortoise has moved another .0164 feet (.005 meters), so he is still behind.

A curious result

It seems that Achilles will never catch the tortoise because as soon as he has moved to where the tortoise was, the tortoise has just inched ahead. Yet this is obviously false.

If the race were run over a course of 328 feet (100 meters), Achilles would run the whole thing in 20 seconds while the tortoise would take 190 seconds to run his 311 feet (95 meters) (taking the 16.4 foot [5 meter] head start into account). Here it is obvious that Achilles wins comfortably. This is the paradox.

GEOMETRY
& TRIGONOMETRY

History of geometry

The word "geometry," the study of things such as points, lines, shapes, and areas, comes from the ancient Greek *geo*, meaning "earth," and *metrein*, meaning "to measure." This etymology demonstrates that this area of mathematics was originally developed to handle everyday concepts such as measuring fields or constructing buildings.

Although the word is Greek in origin, the concepts of geometry were known much earlier. The Babylonians and Sumerians knew about Pythagoras's theorem thousands of years before the Greek philosopher of that name made it popular, and from calculations that appear in the Rhind papyrus (see page 12), we know that the Egyptians were using geometry three and a half millennia ago.

Euclid

Much of our knowledge of the early history of geometry comes from the work of the Greek philosopher Euclid. Euclid lived from around 325–265 BC and taught in Alexandria in Egypt. His major work was the 13-volume series *The Elements*, which brought together proofs of geometrical theorems from earlier mathematicians and added some original work of his own.

Babylonian clay tablet YBC 7289 from the Yale Babylonian Collection, showing evidence of knowledge of Pythagoras's theorem.

A woman teaching geometry in an early Latin translation of Euclid's *Elements*.

FACT

The Indian mathematician Baudhayana lived around 800 BC and was the author of what is known as a Sulba Sutra, which gave details of how religious altars should be constructed. The texts show that the ancient Indians also knew about Pythagoras's theorem.

Basic geometry terms

Points

One of the most basic concepts in geometry, a point indicates a specific location. A point has no size, and can be anywhere in the entire universe.

Point

Line segment

Intersection

Parallel line segments

Lines, intersections, and planes

If you link two points by drawing a rule from one to the other, this is known as a line segment. If you were able to extend the ends of that segment forever in both directions, it would be called a line.

The point where two line segments meet is known as an intersection. Two lines that never meet are said to be parallel.

Imagine you are holding a piece of paper flat on your hand and that it extends in all directions forever (without any wrinkles). That infinitely big, flat surface is known as a plane.

Area

If we now draw a number of intersecting line segments on a sheet of paper to create an enclosed shape, we use the concept of area to define the extent of that shape. Area is usually expressed in square dimensions, such as square feet, miles, or meters. Beyond two dimensions, the concept of area can be used to show the extent of the surface of solid objects such as a ball or pyramid.

Volume

The concept of volume is used to describe how much space is enclosed between the various surfaces of a solid. It is usually expressed in cubic dimensions.

Angles

An angle, which comes from the Latin word *angulus* meaning "corner," represents the amount of rotation between two straight lines as indicated below by the Greek letter theta (θ). The point where these two line segments meet is called a vertex (plural: vertices).

The size of the angle is proportional to the length of the curved line *s* (known as an arc) divided by the radius *r*.

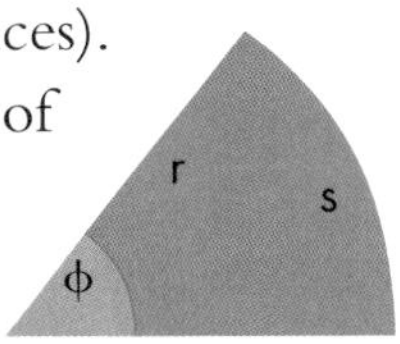

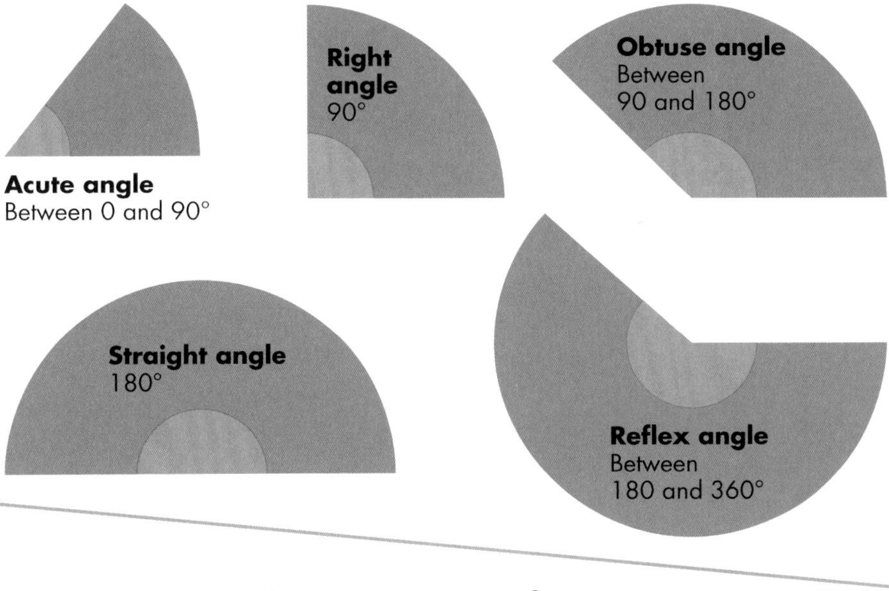

Acute angle
Between 0 and 90°

Right angle
90°

Obtuse angle
Between 90 and 180°

Straight angle
180°

Reflex angle
Between 180 and 360°

Angles are usually given in terms of degrees (represented by the symbol °) and there are 360° in a complete circle. Imagine the angle in the diagram above increasing as the length of the arc (*s*) increases until it's a complete circle.

Different angles have particular names depending on their size, and these are shown in the diagrams above.

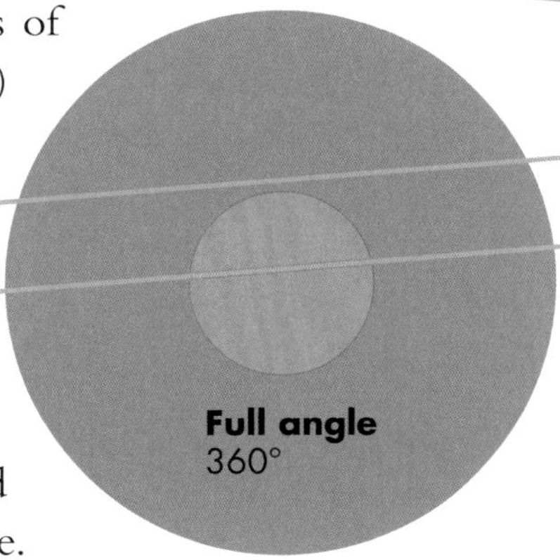

Full angle
360°

Complementary and supplementary angles

Two angles are said to be supplementary if they add up to 180°, while two angles are said to be complementary if they add up to 90°, as show in the examples below.

Supplementary angles

Complementary angles

FACT

The radian is the angle of a pie-shaped segment of a circle where the outer arc and radius are the same length. There are two times π (pi) radians in a complete circle, and so a radian is equal to around 57.3°. Radians are often used in physics and astronomy to describe quantities related to circles or spheres in order to simplify many expressions.

Squares and rectangles

Some of the simplest geometrical shapes we can consider are squares and rectangles. Both of these are part of the class of shapes known as quadrilaterals, which all have four sides (the word comes from the Latin *quadri-* meaning "four" and *latus* meaning "side").

Rectangles and squares represent the group of quadrilaterals whose interior angles are all right angles. A consequence of this is that the opposite sides of squares and rectangles are parallel with each other.

Squares have all their sides the same length, while rectangles have two pairs of sides with equal lengths.

How to calculate area

Imagine you have a piece of paper with 0.39 inches (1 centimeter) wide and high squares marked on it. Draw a rectangle that is 1.57 inches (4 centimeters) wide and 1.18 inches (3 centimeters) high as shown below.

Each 0.39 inch (1 centimeter) square has an area of 0.39 square inches (1 square centimeter); there are 12 of those squares inside the rectangle, and so the rectangle has an area of 4.68 square inches (12 square centimeters). So for any rectangle or square, the area is simply width times height.

Dimension

Put simply, dimension can be defined as magnitude measured in a specific direction. For example, a line could be said to have one dimension (length), while a solid object such as this book has three dimensions: length, width, and depth.

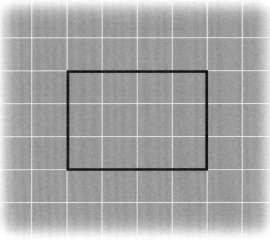

Triangles

Draw three dots on a piece of paper, join them up using straight lines, and you have a triangle. Although the look of your triangle will vary according to where you drew your dots, the angles at the three vertices always add up to 180°.

You can show this by cutting out your triangle, ripping off the three corners, and putting them next to each other against a ruler as shown. This is useful if you know two of the interior angles of a triangle and need to work out the third.

Types of triangle

Triangles can be classified into a number of different types, according to the lengths of their sides. These are given below. (There is also another type: the right triangle, or right-angled triangle, which includes a right angle at one of its vertices.)

Type of triangle	Shape	Description
Scalene		The triangle's sides are all a different length. None of the angles are the same.
Isosceles Note: The dashes show the sides, which have the same length.		Two of the triangle's sides are the same length. Two of the angles inside the triangle are the same.
Equilateral		All three of the triangle's sides are the same length and all three angles are equal to 60°.

Trigonometry

Trigonometry is the mathematical study of triangles and has been around for thousands of years. It is clear that the Egyptians knew of trigonometry when building the pyramids, since trigonometry problems are mentioned in the Rhind papyrus (see page 12). However, it is thought that the Babylonians knew about trigonometry even earlier than the Egyptians.

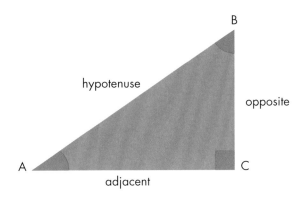

Triangle terminology

The vertices of a triangle are usually labeled with capital letters, such as A, B, and C as shown, and the interior angles at those vertices are often referred to using the same letters. We would then talk of triangle ABC to refer to it.

When we are dealing with right triangles, the longest edge is called the hypotenuse (pronounced "high-pot-en-ooze"). When using trigonometry, we usually know one of the other angles inside the triangle as well. The side next to this known angle is called the adjacent edge, while the remaining side is called the opposite edge for reasons that should be fairly clear. So in the image above, we are assuming that angle A is known.

Pythagoras's theorem

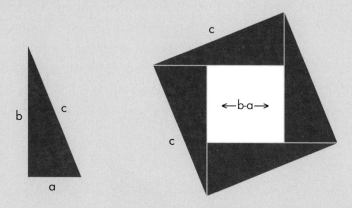

The Greek mathematician Pythagoras is best known for his theorem regarding the lengths of the sides of a triangle (despite the fact that it was actually known well before his time).

The theorem only applies to right-angled triangles and states that if you square the lengths of the two shorter sides of a right triangle and add them together, the number you get is equal to the square of the length of the longest side.

So if we have a right triangle with the two shortest sides measuring 3 cm and 4 cm, then we work out $3^2 = 9$ and $4^2 = 16$ and add them together to get 25. Pythagoras tells us that this is the length of the longest edge squared. 25 is equal to 5^2, so we can say that the longest edge measures 5 cm.

Putting the theory into practice

We know that the area of the blue triangle above is a times b divided by 2 (see box opposite). You can probably see for yourself that the edge of the white square in the middle has a length of b minus a. Therefore the area of the diagram on the right is the area of the four blue triangles added to the area of the square, or:

$$4 \times \frac{a \times b}{2} + (b\text{-}a) \times (b\text{-}a)$$

This can be simplified to:

$$2ab + b^2 - 2ab + a^2$$

or:

$$a^2 + b^2$$

But we also know that the area of the big square (i.e. the combination of the four blue triangles and the white square) is just the two sides multiplied together, c times c, or c^2, so we can write:

$$c^2 = a^2 + b^2$$

This is Pythagoras's theorem in mathematical form. Using letters to stand for numbers in mathematics is called algebra (see chapter 4). What

Gregor Reisch's *Allegory of Arithmetic* (1504), showing Pythagoras with an abacus and the Roman philosopher Ancius Boethius using arabic numerals and mathematical symbols.

we have also done here is proved Pythagoras's theorem no matter what lengths the sides of the triangle are. Doing something this way is called a mathematical proof (see chapter 4).

FACT

There is a simple equation for working out the area of a right triangle. Multiply the lengths of the two shorter sides together and divide by two. So if you have a right triangle with shorter sides measuring 3 cm and 4 cm, the area of the triangle is $3 \times 4 = 12$, divided by 2, which equals 6 square centimeters.

Sines, cosines, and tangents

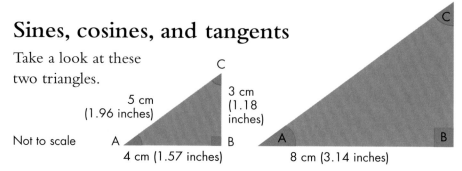

Take a look at these two triangles.

5 cm (1.96 inches)

Not to scale

4 cm (1.57 inches)

3 cm (1.18 inches)

8 cm (3.14 inches)

The second is simply an enlargement of the first one by a factor of two. The angles have all stayed the same, while the length of each adjacent edge has doubled. What do you think the lengths of the hypotenuse and opposite edges are? If you guessed 10 cm (3.93 inches) and 6 cm (2.36 inches) (double the lenghts in the first), you would be correct. What is interesting here is that the ratio of the lengths of the sides (i.e. one divided by the other) has not changed.

A universal truth

In fact, the same is true for any similar triangles as long as the angles remain the same. These values, which are constants for specific angles, have special names in mathematics: sines, cosines, and tangents (sin, cos, and tan are the abbreviations). They are defined as follows:

$$\text{sine of angle A} = \sin A = \frac{\text{opposite}}{\text{hypotenuse}}$$

$$\text{cosine of angle A} = \cos A = \frac{\text{adjacent}}{\text{hypotenuse}}$$

$$\text{tangent of angle A} = \tan A = \frac{\text{opposite}}{\text{adjacent}}$$

If you know one angle and one length, you can use these equations to calculate all the other lengths. You can use the mnemonic soh–cah–toa to remember them, where s stands for sin, c for cos, t for tan, o for opposite, a for adjacent, and h for hypoteneuse.

Sine, cosine, and tangent values for some important angles.

0	0°	30°	45°	60°	90°
Sin	0	0.5	0.707	0.866	1
Cos	1	0.866	0.707	0.5	0
Tan	0	0.577	1	1.732	∞

If you draw graphs of the sine and cosine of angles between 0 and 360°, you get a nice smooth curves of peaks and troughs, while the graph for tangents is much stranger, with the value shooting off to infinity close to 90° and 270°. Note that these graphs then repeat themselves for angles higher than 360°.

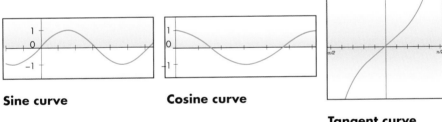

Sine curve   **Cosine curve**

Tangent curve

You will notice that the sine and cosine graphs look similar. In fact, they are identical but shifted across by 90°.

Trigonometric equations

You have a right triangle with the opposite edge measuring 5.77 cm (2.27 inches) and the adjacent edge measuring 10 cm (3.93 inches). What is the size of the angle between the adjacent edge and the hypotenuse? Knowing what you know about the angles in a triangle, can you now work out the angle between the opposite edge and the hypotenuse?

Other important trigonometric functions and equations

Other trigonometrical equations that pop up less frequently are the reciprocal functions to sine, cosine, and tangent, which are known as **secant, cosecant,** and **cotangent,** respectively.

Reciprocal function	Abbreviation	Calculation
secant of angle A	= sec A	= $\dfrac{\text{hypotenuse}}{\text{adjacent}}$
cosecant of angle A	= csc A	= $\dfrac{\text{hypotenuse}}{\text{opposite}}$
cotangent of angle A	= cot A	= $\dfrac{\text{adjacent}}{\text{opposite}}$

You will also sometimes see squared trigonometrical functions. They are written in a particular way to avoid confusion with finding the sine or cosine of A^2:

$$\sin^2 A = \sin A \times \sin A$$

$$\cos^2 A = \cos A \times \cos A$$

Trigonometrical identities

Imagine you have two angles, one called A equal to 30° and another called B equal to 45°. Using your knowledge of trigonometrical identities and the values of sin and cos for these angles from the table earlier in this chapter, can you work out the values of sin 15°, cos 15°, sin 75°, and cos 75°?

Trigonometric identities

We know from Pythagoras's theorem that for a triangle with hypotenuse of length c, adjacent edge of length a, and opposite edge of length b:

$$a^2 + b^2 = c^2 \quad \text{(equation 1)}$$

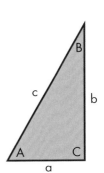

The SOHCAHTOA rules tell us that:

$$\sin A = \frac{b}{c} \quad \text{and} \quad \cos A = \frac{a}{c}$$

If we square both of these we get:

$$\sin^2 A = \frac{b^2}{c^2} \quad \text{and} \quad \cos^2 A = \frac{a^2}{c^2}$$

and multiplying both sides of these two equations by c^2 we get:

$$c^2 \sin^2 A = b^2 \quad \text{(equation 2)}$$

and

$$c^2 \cos^2 A = a^2 \quad \text{(equation 3)}$$

Combining equations 1, 2, and 3 we get:

$$c^2 \cos^2 A + c^2 \sin^2 A = a^2 + b^2 = c^2$$

and dividing this by c^2 we arrive at:

$$\cos^2 A + \sin^2 A = 1$$

This is known as the Pythagorean identity and is useful for simplifying complex trigonometrical equations. Because we did not actually say what the angle A was, we can use it for any angle A.

How tall is the Eiffel Tower?

Imagine that we want to figure out the height of the Eiffel Tower in Paris, France. Let's assume that we are standing 173 meters (567 feet) from the center of the base of the tower, looking up at the top, our head tilted at an angle of approximately 60°. If we impose a right-angled triangle on top of these positions, we find that the right angle slots nicely into the base of the tower. Let's call the angle at the base of the tower angle B. We are standing at angle A and the top of the tower is home to angle C. From here we draw up the following equation:

$$\frac{BC}{AB} = \tan 60°$$

To find BC (the height of the tower), we arrange the equation to give the following:

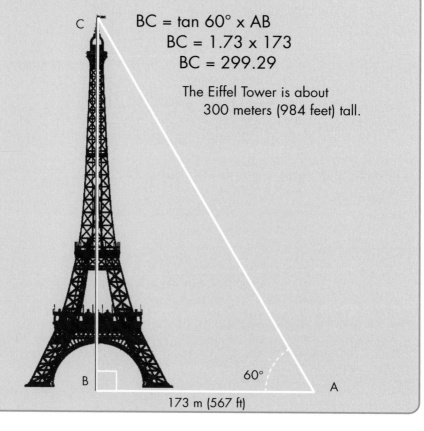

$$BC = \tan 60° \times AB$$
$$BC = 1.73 \times 173$$
$$BC = 299.29$$

The Eiffel Tower is about 300 meters (984 feet) tall.

60°

173 m (567 ft)

Circles

The distance from the center of a circle to its edge is the radius. The diameter is twice the radius, and the circumference is the distance around the edge. The number π (pi) is defined as the circumference divided by the diameter. For a circle of radius r the circumference $= 2\pi r$. The area of a circle can be calculated using the formula πr^2.

Circular elements

Some other elements of a circle are arcs, chords, sectors, and segments. An arc is just a continuous section of the circumference, as indicated in blue. The length of an arc is closely related to the circumference of the circle and the angle θ between the two radii defining the arc by the following equation:

$$\frac{\theta}{360°} = \frac{\text{length of arc}}{\text{circumference}}$$

You will sometimes hear mathematicians talk about minor and major arcs. In our diagram the minor arc is in blue. The remaining edge of the circumference is called the major arc. If we join together the two points where the radii touch the arc with a straight line, we get what is known as a chord. The area between a chord and the minor arc is known as a segment. The wedge between two radii, meanwhile, is known as a sector.

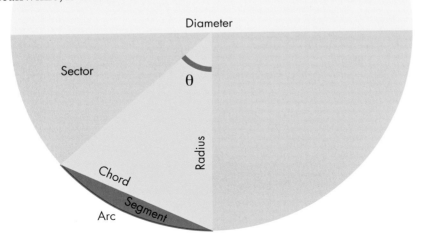

FACT
The interior angle of a regular polygon is given by
$$\frac{180n - 360°}{n}$$
where n is the number of sides.

Other quadrilaterals

Two other four-sided shapes that you might encounter are the parallelogram and the rhombus. The opposite sides of a parallelogram are parallel (little surprise there) and the same length. However, the

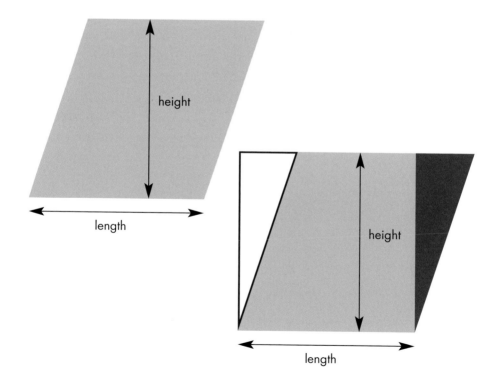

sides do not all have to be the same length. Where the four sides are equal, the shape is known as a rhombus.

The area of a parallelogram is equal to its base length multiplied by its height. You can see this by making one in paper and cutting the end off as shown on the oppposite page in red. The piece fits on the other side in the space with a red outline, creating a rectangle, and we know that the area of a rectangle is its two sides multiplied together.

Polygons

All of the two-dimensional shapes we have discussed already, such as squares, circles, parallelograms, and triangles, come under the general heading of polygons. These are shapes that are totally enclosed by straight line segments. Regular polygons have edges that are all the same length, and all of their interior angles are also the same.

Regular shapes with more sides

Pentagon
Sides: 5
Interior angles: 108°

Hexagon
Sides: 6
Interior angles: 120°

Heptagon
Sides: 7
Interior angles: 128.571°

Octagon
Sides: 8
Interior angles: 135°

Decagon
Sides: 10
Interior angles: 144°

Dodecagon
Sides: 12
Interior angles: 150°

Three-dimensional objects

Now that we have looked at all of the major two-dimensional (sometimes called planar) objects, we turn our attention to three dimensions.

Cubes and cuboids

We start with the cube, a three-dimensional object with six square faces, each of which meets two other faces at each vertex. A cuboid is similar, but its faces are all rectangles, rather than squares.

When working with three-dimensional shapes, we often want to work out their volume, the amount of space enclosed by the object. The first step is to define a unit of measurement. The volume of a cube measuring 1 cm × 1 cm × 1 cm is called a cubic centimeter or 1 cm³.

Now imagine we have a cube of butter measuring 10 cm × 10 cm × 10 cm and we chop it up into smaller cubes, measuring 1cm³ each. If we count them out, we end up with 1,000 of these smaller cubes; this is clearly the number you get when you multiply 10 by 10 by 10.

In fact, the volume of any cuboid, including cubes, is just the length, multiplied by the width, multiplied by the height, or:

$$V = l \times w \times h$$

So a cuboid measuring 10 × 20 × 30 cm would have a volume of 6,000 cubic centimeters or 6,000 cm³.

The other thing we often want to do with a three-dimensional object is to measure its surface area. A cuboid with edges of length *l*, *w*, and *h* has six rectangular faces. We know that the area of a rectangle is its two sides multiplied together, and it is easy to see that the total surface area should be:

$$\text{Area of cuboid} = 2 \times (lw + lh + hw)$$

For a cube with an edge of length a, these two equations get simpler:

$$\text{Volume of a cube} = a^3$$

$$\text{Surface area of a cube} = 6a^2$$

Spheres

A sphere is the three-dimensional counterpart of the circle. They are defined as objects whose surface lies a fixed distance away from a central point. That fixed distance, in the case of both circles and spheres, is known as the radius, which we usually denote with the letter r.

The surface area and volume of a sphere are given by the following equations:

$$\text{Volume of a sphere} = \frac{4\pi r^3}{3}$$

$$\text{Surface area of a sphere} = 4\pi r^2$$

Surface area and volume of a cuboid

You have bought your niece a present of a set of 27 building blocks. Each block measures 7 cm x 3 cm x 2 cm. The blocks fit snugly into a box with nine blocks per layer (arranged in a three by three formation). What is the volume of an individual block?

Give an estimate of the volume of the box in which they fit (assuming the box is of negligible thickness). What is the area of the smallest piece of wrapping paper you could use to gift wrap it?

Pyramids, prisms, and cylinders

Mathematicians often look at other important three-dimensional solids. Their areas and volumes are calculated as follows:

Pyramid
Faces: Five (square base of edge length a, plus four triangular sides of slant length s; pyramid height h).
Volume = $\frac{1}{3} a^2 h$
Surface area = $2as + a^2$

Prism
Faces: Depends. A prism is an object that has the same cross-section (of area A) along its entire length, l.
Volume = Al
Surface area: Depends on type of prism

Cylinder
Faces: Three (circular base and top with radius r and a curved side of height h).
Volume = $\pi r^2 h$
Surface area = $2\pi rh + 2\pi r^2$

Cone
Faces: Two (circular base of radius r and a curved edge with a slanted edge of length s).
Volume = $\frac{1}{3} \pi r^2 h$
Surface area = $\pi rs + \pi r^2$

Polyhedrons

There are also other regular solids that we might come across with a greater number of faces, shown in the diagrams below. Solids with flat faces and straight edges are collectively known as polyhedrons.

Tetrahedron
Four triangular faces.

Octahedron
Eight triangular faces.

Dodecahedron
12 faces, usually pentagonal.

Icosahedron
20 triangular faces.

Cones

The Greeks were fascinated with cones. Around 400 BC, it is known that Plato and his followers were studying cones and their properties. Euclid (see page 40) was also an authority on the geometry of cones. It is believed he wrote four books on the subject, although we know this only by reference to them in other literature from years later.

Conic sections

However, it is Apollonius of Perga, who lived from around 262 BC –190 BC, who is remembered most for his work with cones and their properties.

His biggest contribution was in the area of so-called conic sections, which are the curves you get when you slice through a cone in different directions. Slicing at different angles produces different types of curves, as shown in the images below.

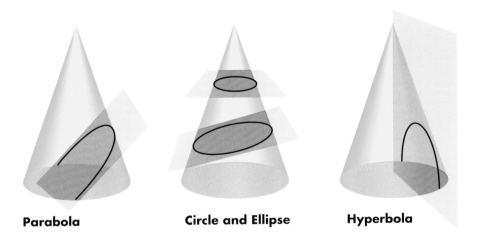

Parabola   **Circle and Ellipse**   **Hyperbola**

Ellipses

The ellipse is the mathematical name for a squashed circle or oval, although it does have a more mathematical definition. It is the curve that is traced out where the sum of the distances from two other fixed points (which are called the foci) is a constant.

The two red dots labeled F1 and F2 below are the foci.

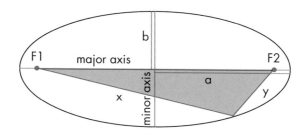

Major and minor axes

The distance from edge to edge passing through both F1 and F2 is called the major axis, while the distance from top to bottom passing through the center is known as the minor axis.

In our definition, the sum of the distances (i.e. $x + y$ in our diagram) from the two foci is a constant. In fact, the following applies:

$$\text{Major axis: } 2a = x + y$$

We can also calculate the length of the minor axis. If the distance between F1 and F2 is called f, then:

$$\text{Minor axis: } 2b = \sqrt{(x + y)2 - f2}$$

The amount the ellipse is squashed is given by the eccentricity, e, which equals:

$$\sqrt{\frac{a^2 - b^2}{a^2}}$$

The other two things you might want to know about an ellipse are its area and the perimeter. The area is given by the following:

Area of an ellipse = πab

The length of the perimeter of an ellipse is extremely difficult to calculate and requires the use of an infinite series.

However, you can get an approximate value for the perimeter p from the following formula:

$$p \approx 2\pi \sqrt{\frac{a^2 - b^2}{a^2}}$$

Planetary orbits

All the planets follow elliptical paths in their orbits. However, these orbits are only slightly more squashed than a circle, which is why it took until 1609 for a scientist named Johannes Kepler to realize this. He showed that planets do not follow circular orbits as had long been assumed, but elliptical paths, with the Sun at one of the two foci.

Sir Isaac Newton went on to show that orbits can be any type of conic section. However, an object with a parabolic or hyperbolic orbit – which some comets have – would pass the Sun once and then leave the solar system.

The study of ellipses is widely used in astronomy.

Parabolas

A parabola is the curve you get where the distance from a fixed point (known as the focus) and a fixed line (known as the directrix) is a constant.

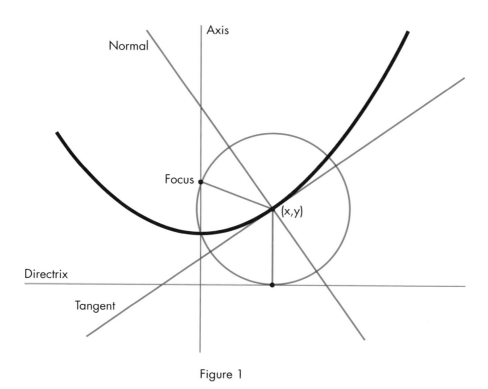

Figure 1

What this means is that in Figure 1, the two blue lines are always the same length no matter where you move on the curve. The equation of a standard parabola has the form:

$$y = ax^2 + bx + c$$

A graph of the function $y = x^2$ is therefore a very simple parabola. Comets traveling around the Sun follow parabolic paths.

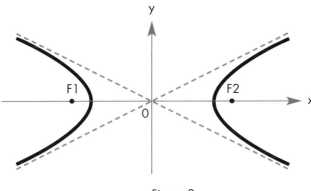

Figure 2

Hyperbolas

A hyperbola is two curves that are the mirror image of each other. In mathematics, it is the curves made up from the points where the difference between the distance to two foci (F1 and F2 in the diagram above) is a constant.

The dotted lines are called the asymptotes and the curves get closer and closer to these lines but never actually meet them. A hyperbola typically can be represented by the equation:

$$\frac{x^2}{a^2} - \frac{y^2}{b^2} = 1$$

where a and b are related to the distance between the two foci.

FACT

Parabolas crop up regularly in everyday life. One use is in car headlamps, where the rear reflector is usually parabolic. The lamp itself sits at the focus of the parabola. This arrangement means that light emerging from the lamp hits the reflector and then emerges in a parallel beam.

Coordinates

Coordinates define the positions of points and objects relative to another location. The graph below shows two lines at right angles to each other – one vertical (the y axis) and one horizontal (the x axis).

The point where the two lines meet is called the origin. The two axes are numbered in steps of one in both directions.

Plotting a location

Look at the red dot on the diagram, and trace your finger from the dot to the y axis and then the x axis to find out its location. We say this point has an x coordinate of 6 and a y coordinate of 8.

Coordinates are usually written in brackets, with the x coordinate first and the y coordinate second, separated by a comma. In this case the coordinates are (6,8). The blue dot is at the coordinates (9,5).

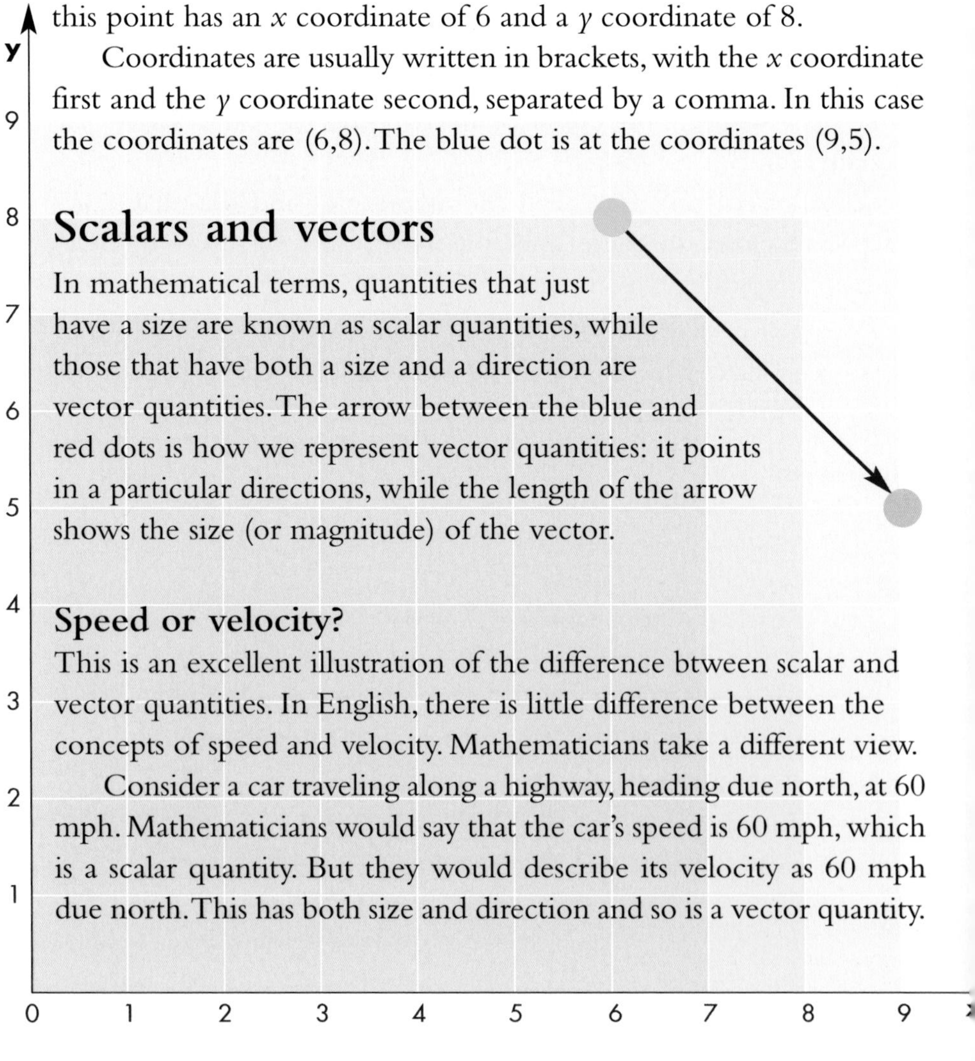

Scalars and vectors

In mathematical terms, quantities that just have a size are known as scalar quantities, while those that have both a size and a direction are vector quantities. The arrow between the blue and red dots is how we represent vector quantities: it points in a particular directions, while the length of the arrow shows the size (or magnitude) of the vector.

Speed or velocity?

This is an excellent illustration of the difference btween scalar and vector quantities. In English, there is little difference between the concepts of speed and velocity. Mathematicians take a different view.

Consider a car traveling along a highway, heading due north, at 60 mph. Mathematicians would say that the car's speed is 60 mph, which is a scalar quantity. But they would describe its velocity as 60 mph due north. This has both size and direction and so is a vector quantity.

Symmetry

The harmonious appearance of objects that we know as symmetry comes from the mathematical operations of reflection, translation, and rotation. Repeated patterns such as you see in brick walls, wallpaper, and carpets are the result of the mathematical operation of translation, with a single image translated and repeated many times over and over again.

This pattern is created by the repeated translation of a single symmetrical shape.

Lie groups

Introduced by the 19th-century Norwegian mathematician Sophus Lie, these groups help us to understand symmetry from a mathematical perspective.

Every symmetrical object, such as a sphere, can be represented by a Lie group (a set containing numbers and a mathematical operation on those numbers) in three dimensions.

However mathematicians are not limited to just three dimensions. An international team of 18 mathematicians has mapped out a structure for a Lie group known as E8, which takes symmetry to 248 dimensions. If you wrote down all the equations dictating the structure of E8, they would cover an area the size of Manhattan.

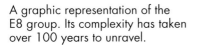

A graphic representation of the E8 group. Its complexity has taken over 100 years to unravel.

Topology

Topology first became popular in the 18th century, when the Swiss mathematician Leonhard Euler published a paper on a problem known as the Seven Bridges of Königsberg.

The city of Königsberg was set on both sides of a river, including two islands in between the forks of the river. The islands were connected to the mainland areas and to each other by seven bridges. The challenge was to find a way to cross each bridge only once in a single journey, returning to your original point. Starting from Point A, can you make your way around the city without retracing your steps?

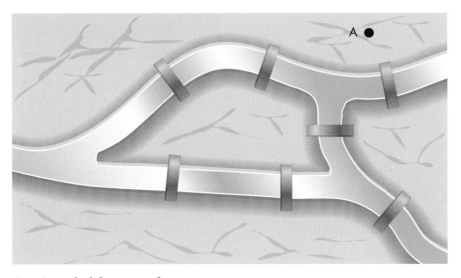

An insoluble puzzle

The solution to the problem was never found, as it is impossible to cross all bridges just once in a single journey. Euler proved that it was impossible by simplifying the map of the city into a graph, with each piece of land represented by a dot (known as a vertex or node) and each bridge represented by a line (known as an edge).

For mathematicians, topology is the study of how complex situations can be simplified into these graphs. Fundamentally, the general shape of these graphs is not important, just the way that the nodes and edges interact with each other.

ESSENTIAL
ALGEBRA

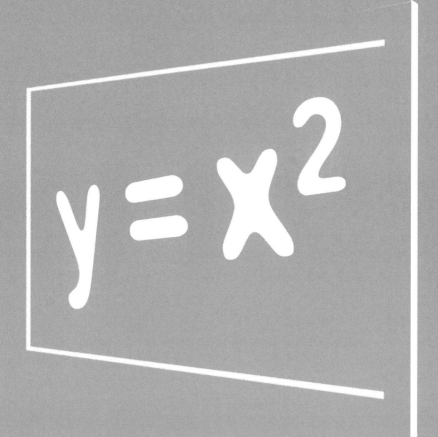

The use of symbols

At the heart of algebra is the idea of replacing numbers in equations with symbols, usually letters of the alphabet. This allows us to formulate general rules about numbers and also enables us to work out unknown quantities.

Variables and constants

Take a look at the simple formula $y = x + 2$. The letters x and y in this equation are called variables because they can take on any value. For example, if $x = 2$, then $y = 4$ and if $x = 5$, then $y = 7$ and so on for any value we choose for x.

Constants are numbers that have a fixed value. In the formula above, the number 2 is a constant. However, constants may also be represented by letters. Constants are used in algebraic equations (often called expressions) such as $y = ax^2 + bx + c$. Here the letters a, b, and c represent constant values.

The expression $y = ax^2 + bx + c$ is what mathematicians call a polynomial: the sum of several individual terms (i.e. ax^2, bx, and c) that involve constants and variables.

Algebra's origins

The word "algebra" comes from the Arabic *al-jabr*, although there is some uncertainty about exactly what it means – some say it means "reunion," some say "completion," while others say it translates better as "balancing."

What's in a name?

As a word, algebra first appears in the title of the book *Hisab al-jabr w'al-muqabala* by the Persian mathematician Muhammad ibn Musa al-Khwarizmi. The book, whose title can be loosely translated as *The Comprehensive Book of Calculation by Balance and Opposition*, was published in around 820 AD and looked at how to handle various types of algebraic expression.

Although he gave us the word "algebra," al-Khwarizmi did not use letters to represent variables. Instead, he wrote everything out in words, using the word "squares" (to represent x^2), "roots" (to represent terms such as $5x$), and "shay" (which translates as "thing") to mean an unknown variable. Some scholars believe that when his work was translated into Spanish, "shay" was written as "xay" and that our use of x comes from an abbreviation of this.

Earlier sources

Despite al-Khwarizmi's influence, algebraic concepts were known before this period in history.

This Soviet stamp was issued in 1983 to commemorate al-Khwarizmi's (approximate) 1200th anniversary.

The Rhind papyrus (see page 12) reveals that the Egyptians knew about linear equations. The Chinese also understood algebraic ideas. One of the chapters of *The Nine Chapters on the Mathematical Art* (see page 13) discusses how to solve simultaneous equations using a technique called fang cheng, based on counting rods.

FACT

Pythagoras's theorem (see pages 48–9) is an algebraic expression that was known well before al-Khwarizmi wrote his book. However, like other algebraic ideas, it was often handled through geometry.

Graphs

A graph – a pictorial representation of an equation – is often a very good way to get an insight into what an equation actually means.

We can define a single point in space using x and y coordinates (see page 66) relative to some point of origin that we define. A graph allows us to take this idea further, by letting us show all of the points whose values of x and y fit some relationship that we can represent with an equation.

Using a graph

You can use graphs to solve equations. Say you want to find out the values of x that satisfy the equation $x^3 + 5x = x^2 + 6$. If you plot the graphs $y = x^3 + 5x$ and $y = x^2 + 6$ and see where they intersect, that will give you the solution to the original equation.

Graphs don't have to start at zero. Often you'll see graphs with negative numbers. The axes still cross at $x = 0$, $y = 0$, but they can start from anywhere, and can go up in multiples as well as ones.

GREEN LINE
This plots the graph of the equation $y = x^2 + 1$. To check this, choose a value of x such as 1, square it and then add 1. This gives the answer 2. So when $x = 1$, $x^2 + 1 = 2$. So we plot a point on the graph at $y = 2$, $x = 1$.

BLUE LINE
This shows every point on the graph where y has the same value as x $(y = x)$.

RED LINE
This simple line shows every point on the graph where $y = 2$.

Y

X

-9 -8 -7 -6 -5 -4 -3 -2 -1   0   1   2   3   4   5   6   7   8   9 10

Solving cubic equations using graphs

The Persian mathematician Ghiyath al-Din Abu'l-Fath Umar ibn Ibrahim al-Nisaburi al-Khayyami is better known by the name of Omar Khayyam. He lived from 1048 to around 1131 AD in Nishapur, now in Iran.

His major work on algebra was the *Treatise on Demonstrations of Problems of Algebra*, published in 1070. In it, Khayyam proposed techniques for solving cubic equations (i.e. those where the highest order terms involve x^3) by drawing graphs of two different conic sections (see page 61) and looking at the values of x where they intersect.

Khayyam also became renowned as a poet after Edward Fitzgerald's beautifully illustrated 19th-century English translation, *The Rubaiyat of Omar Khayyam*.

FACT

The right-hand side of an equation such as $y = 8x^2 + 7x^2 + 6x + 5x + 7$ is known as an algebraic expression, and each individual element added together as a term of that expression. Terms that have the same power of x are known as like terms (e.g. $8x^2$ and $7x^2$), while those with different powers (e.g. $8x^2$ and $5x$) are non-like terms. The numbers 8, 7, 6, and 5 are what we call the coefficients of those particular terms.

Types of equations

The highest power of x (or whatever variable it is we are trying to discover) in any equation tells us what sort of equation it is.

Linear

If an equation only has terms containing x (i.e. x to the power 1) and nothing with higher powers, such as x^2 or x^3, then we call the equation linear. A typical linear equation looks like this:

$$ax + by + c = 0$$

where x and y are variables and a, b, and c are constants.

If you draw a graph of such an equation, you get a straight line, hence the name "linear."

Quadratic

An equation whose highest term contains x^2 (i.e. there are no terms containing x^3 or anything higher) is called a quadratic equation. A typical quadratic equation looks like this:

$$ax^2 + bx + c = 0$$

where a, b, and c are constants and a is not equal to zero (if it were, then the equation would be linear).

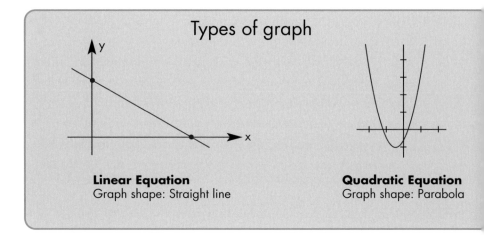

Types of graph

Linear Equation
Graph shape: Straight line

Quadratic Equation
Graph shape: Parabola

Higher polynomials

Equations with x^3 as the highest power term are called cubic, those with x^4 quartic, and those with x^5 quintic. There is no limit to the power of x in polynomial terms, but those beyond those listed above are rarely encountered and are incredibly difficult to solve.

Terms in x (such as $5x$), x^2 (such as $7x^2$), and x^3 (such as $8x^3$) have special names: they are called monomials, binomials, and trinomials, respectively.

The shapes of graphs

Graphs of different types of equations have different shapes. Some of the most common are given below.

Critical points

A critical point is a local peak or trough in the curve of a graph. More precisely, it is any point where the tangent line to the curve is horizontal, i.e. the slope of the tangent is zero.

If you take a look at the graph shapes below, you will notice that the number of critical points is always one less than the highest power term in the equation, i.e. a quadratic equation with its highest term in x^2 has one critical point.

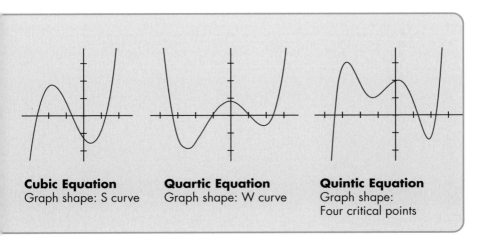

Cubic Equation
Graph shape: S curve

Quartic Equation
Graph shape: W curve

Quintic Equation
Graph shape:
Four critical points

Graphical methods of solution

Until al-Khwarizmi found a way of solving quadratic equations by algebraic means, the only way to do so was using geometry or graphical methods.

Imagine that we want to solve the quadratic equation $x^2 - 8x + 15 = 0$, i.e. we want to find out the values of x which, when plugged into the left-hand side of the equation, give an answer of zero.

Now we want to find out the points where $x^2 - 8x + 15 = 0$. Look at the graph below: there are two points where the function (i.e. the value on the y-axis) equals zero, indicated by the two blue dots. All we need to do is read off the values on the x-axis that correspond to those points, which are $x = 3$ and $x = 5$, to find our solutions. (Why not plug them into the equation to check?)

Graphical methods can also be used to solve more difficult problems, such as cubic equations (see page 75).

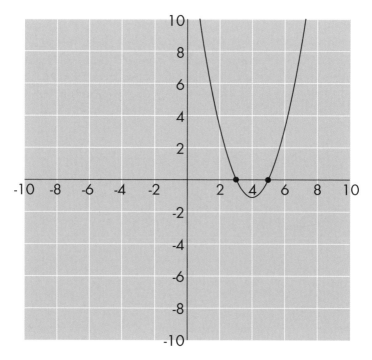

Graph plotting equation $y = x^2 - 8x + 15$

Factors

The factors of a number n are those numbers that can be exactly divided into n. The factors of 12 are 1, 2, 3, 4, and 6 (it is traditional not to include the number itself among its list of factors).

Expanding brackets

Often in algebra, you will see something like the following:

$$a(x + 2y) = 0$$

i.e. a constant multiplied by something inside a pair of brackets. It is effectively a shorthand way of writing something else. We can see what the longhand version was by expanding the brackets, which means multiplying the whole thing out.

We do this by multiplying each thing inside the brackets by the thing outside it and adding them together. In our case, this gives us:

$$ax + 2ay = 0$$

[Note: As multiplication is commutative (see chapter 2), we can write out the terms 2, a, and y in any order without changing the answer of the multiplication. It is usual to write any constant number (such as the 2) first.]

The two equations above mean the same thing, but the first one allows us to avoid repeating the a in the second term. This particular example won't save much ink, but it can make a huge difference. You can also expand more complex equations such as:

$$(x - 5)(x + 4) = 0$$

After multiplying everything in one bracket by everything in the other and adding them together, we get

$$x^2 - 5x - 20 + 4x = 0$$

Collecting the like terms in x together gives us

$$x^2 - x - 20 = 0$$

Expanding the brackets

Using the step-by-step process set out on the previous spread, can you expand the brackets in the following equations:

$$3(y - 5) = 0 \qquad\qquad (x - 5)(x - 7) = 0$$

(Hint: Remember to multiply everything in one bracket by everything in the other, and add them together)

How does this help? Remember that the factors of a number n are those numbers that you multiply together to get n. Similarly, since

$$(x - 5)(x + 4) \ \text{ and } \ x^2 - x - 20$$

are two ways of writing the same thing, we can say that $(x - 5)$ and $(x + 4)$ are factors of $x^2 - x - 20$. This may sound like cheating but, after all, the letters in expressions are just replacements for numbers. Take a look back at the original equation:

$$(x - 5)(x + 4) = 0$$

It says that we are multiplying two things, $(x - 5)$ and $(x + 4)$, together, and the answer is zero. The only way to get an answer of zero when multiplying two numbers together is if one or the other number is itself zero.

In this case, it means that either $x - 5 = 0$ or $x + 4 = 0$. Working out these simple linear equations, we can find that x is either equal to 5 or –4. Or to put it more plainly, 5 and –4 are the two values of x that satisfy the above equation.

Factorization

The process outlined above becomes more powerful when you do it in reverse. This is called factorization, the process of finding the factors of an equation.

Rearranging equations

When faced with an equation, it can sometimes be useful to rearrange it to make it easier to handle. Take this equation:

$$2y^2 + 3y - 3 = \tfrac{1}{2}y^2 + 4/y + 4$$

We can now perform various mathematical operations on it, such as adding or dividing (although not by zero or anything that equals zero) by other constants or variables, to make it easier to understand and to solve using standard methods.

Whatever we do, the most important thing to remember is that the same operation must be done on both sides of the equation.

First, let us make sure all the terms with the same power of y are collected together, i.e. all the terms with y^2 in, for example. Subtracting $\tfrac{1}{2} y^2$ removes the first term of the right-hand side and makes the $2y^2$ on the left into $\tfrac{3}{2} y^2$. We can also subtract 4 from both sides to get rid of the last term on the right-hand side and make the -4 on the left into -7. This leaves us with:

$$\tfrac{3}{2}y^2 + 3y - 7 = 4/y$$

Now look at the remaining term on the right-hand side. It is usually easier not to have to deal with terms such as $1/y$ or $1/y^2$. We can get around this by multiplying both sides by y. (Remember that you have to multiply every term on the left by y.) This gives:

$$\tfrac{3}{2}y^3 + 3y^2 - 7y = 4$$

It is also usually easier to have an equation where all of the terms are on the left by setting the right side to zero. We can do this by subtracting 4 from both sides to give:

$$\tfrac{3}{2}y^3 + 3y^2 - 7y - 4 = 0$$

Finally, we multiply the whole thing by 2 to avoid the fraction in the first term:

$$3y^3 + 6y^2 - 14y - 8 = 0$$

Solving linear equations

Linear equations tend to have the general form $ax + by + c = 0$. The simplest linear equations are those where the constant b is zero, i.e. there is no term containing y. For example:

$$5x - 15 = 0$$

Note that our constant c is actually subtracted in this instance, which seems to make it different from the general form above. It doesn't matter. Subtracting can be treated the same as if you are adding a negative number, i.e. $5x - 15$ is the same as $5x + (-15)$.

To solve the equation, we can use the techniques set out in "Rearranging equations" (see page 79). We want to find out the value of x that satisfies the equation. The way to do this is by rearranging to get x on its own on one side of the equation.

So, for our example, we can add 15 to both sides to get:

$$5x = 15$$

Dividing both sides by 5 gives us the answer: $x = 3$

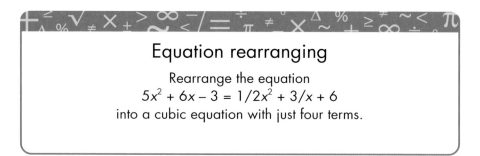

Equation rearranging

Rearrange the equation
$$5x^2 + 6x - 3 = 1/2x^2 + 3/x + 6$$
into a cubic equation with just four terms.

Slope

The slope of the graph of a linear equation is a measure of the angle that the line makes with the horizontal.

To calculate the slope, which is usually denoted by the letter m, you need the coordinates of two points on the line, which we shall call (x_1, y_1) and (x_2, y_2).

[Note: It is common practice to use subscript numbers to denote different points.]

The slope of the line is then given by the equation:

$$m = \frac{y_2 - y_1}{x_2 - x_1}$$

You can pick any two points on the line to do this, since the slope is unchanged everywhere on the line.

In fact for any line given by a formula that looks like $y = ax + c$, where a and c are constants, the slope is equal to the constant a.

You can see from this that the slope of a horizontal line, such as $y = 2$, is equal to zero. That is because you can write $y = 2$ as $y = 0x + 2$.

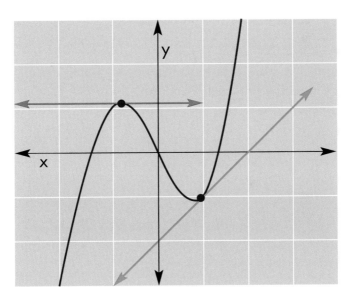

Tangent lines at different points of a graph.

Tangents

The tangent (or tangent line) is the straight line that just touches the curve at any particular point (see green line in the graph right). Each point on the curve has a different tangent. At the first peak the tangent line is horizontal and shown in red.

We can calculate the slope of the tangent line in the same way as the slope of the line (see above.)

Simultaneous equations

Imagine you have two friends who have just been to the coffee shop. One says they have bought a cappuccino and two muffins at a cost of $7. The other says they got three cappuccinos and three muffins for $15. How much would a single cappuccino and a single muffin cost?

The answer is that a cappuccino costs $3 and a muffin $2. You could work it out by trial and error, but it's an algebraic problem. Let's say the cost of a cappuccino is x and the cost of a muffin is y. Writing out the two statements in algebraic form we get:

$$x + 2y = 7 \text{ (equation 1)}$$

and:

$$3x + 3y = 15 \text{ (equation 2)}$$

Solving the puzzle

A boy says: Two years ago, my father was four times as old as me. His father says: In three years, I will be three times my son's age. How old are the father and son?
{Hint: Use the variables B and F to represent their ages}

We can then play with these equations to eliminate either x or y and thus to solve the problem.

Let's start by multiplying every term in the whole of the first equation by 3. Since we are doing the same to both sides, this does not change the validity of the statement. This gives us:

$$3x + 6y = 21 \text{ (equation 3)}$$

Now both equations have a term $3x$ in them.

Next we do something clever: we subtract equation 2 from equation 3. We do this by subtracting every term in equation 2 from every corresponding term in equation 3 as follows:

$$3x - 3x + 6y - 3y = 21 - 15$$

This can be tidied up to give us:

$$3y = 6$$

Dividing this throughout by 3 gives us $y = 2$, i.e. the cost of the muffin is $2. Now we can substitute this value into either equation 1 or 2 to find out that $x = 3$, i.e. the cost of the cappuccino is $3.

Solving simultaneous equations

The chart below sets out the process for solving simultaneous equations. Note that, for this technique to work, you need the same number of equations as there are unknown variables.

Four steps to solving simultaneous equations

The procedure to follow is always the same:

1. Multiply one of the equations by a number so that one of the unknown terms in the equation is the same as in the other equation.
2. Subtract one equation from the other so that one of the unknown terms disappears.
3. Calculate the remaining unknown variable.
4. Use this value in one of the original equations to find the other unknown.

Solving quadratic equations

Sometimes it isn't possible to factorize an equation, in which case we need another method. Take the following example:

$$x^2 + 4x - 2 = 0$$

It is impossible to factorize, but we can do it another way.

Take the constant associated with the x term, in this case 4, divide it by 2 to get 2, then square it to get 4. Add that to both sides:

$$x^2 + 4x + 4 - 2 = 4$$

The first three terms can then be rewritten as $(x + 2)^2$:

$$(x + 2)^2 - 2 = 4$$

or we can rearrange it by adding 2 to both sides:

$$(x + 2)^2 = 6$$

Taking the square root of both sides gives us:

$$x + 2 = \pm\sqrt{6}$$

A final rearrangement gives us:

$$x = \pm\sqrt{6} - 2$$

[Note: The $\pm$ symbol means that the root can be positive or negative, since we don't know which it is. There are always two solutions to a quadratic equation.]

Tap this into a calculator and you find that the two values of x that satisfy this equation are 0.449 and –4.449 (to three decimal places).

FACT

Quadratic equations are used extensively in many industries, but they tend to be masked by the increasing use of computers to solve design and engineering problems. One area where quadratic equations are used is in calculating the safe braking distance for cars.

Inequalities

Inequalities are mathematical statements about the relative sizes of quantities. They use the following symbols:

$<$	$>$	$\leqslant$	$\geqslant$	$\neq$
Less than	Greater than	Less than or equal to	Greater than or equal to	Not equal to

A simple inequality might be $x > 5$, a statement that says whatever the value of x is, it is always greater than 5.

The rules of inequalities

As with equations, we can perform mathematical operations on them by applying them to both sides of the inequality.

Applicable rules for mathematical operations on inequalities
• If $x > y$ and $y > z$, then we know that $x > z$
• If $x < y$ and $y < z$, then we know that $x < z$
• If $x > y$ and $y = z$, then we know that $x > z$
• If $x < y$ and $y = z$, then we know that $x < z$
• If $x < y$, then we can say $x + z < y + z$ and $x - z < y - z$
• If $x > y$, then we can say $x + z > y + z$ and $x - z > y - z$
• If $x < y$ and z is positive, then we can say $xz < yz$
• If $x < y$ and z is negative, then we can say $xz > yz$
• If $x < y$ then $-x > -y$
• If $x > y$ then $-x < -y$
• If $x > y$ then $1/x < 1/y$
• If $x < y$ then $1/x > 1/y$
where x, y, and z are all real numbers.

Using inequalities

We can use our known inequalities and the rules for performing operations on them to make calculations. Let's say you have $20. You want to buy a scarf, which costs $8, and some books, which cost $3.70 each. How many books can you buy?

We can write this as an inequality, where b is the number of books:

$$5 + 3.70b < 20$$

We now subtract 5 from each side:

$$3.70b < 15$$

and then divide both sides by 3.70:

$$b < 4.05 \text{ (to 2 decimal places)}$$

i.e. the number of books has to be less than 4.05, therefore we can buy a maximum of four.

The cookie question

You have $11 and want to buy some cookies. Chocolate cookies cost 20 cents each, while raisin cookies cost 15 cents. You prefer chocolate cookies but want to buy some of both types. In fact, you want exactly twice as many chocolate as raisin ones. What is the most cookies you could buy in total?

[Hint: Use variables to represent the two different cookie types. Write out the two requirements as inequalities then solve them.]

Proof

Mathematicians like to be able to prove what they do. By proving something, you show that an equation can be used no matter what values you plug into it.

The Greek mathematician Euclid was the first to lay down formal rules for mathematical proofs. His system was to make some initial assumptions and then, by using logic, show that if these are true, then some derived conclusion is also true. You are also allowed to introduce any previously proven results along the way.

There are three main ways to prove things mathematically:

- **Direct proof**
  This is where you use already proven ideas and assumptions to prove a statement. You could use a direct proof to prove that adding one to an even number gives an odd number by starting off with the knowledge that an even number e is defined as being twice some other whole number.
- **Proof by induction**
  With a proof by induction, you can prove that some basic idea is true directly before going on to prove that all following ideas in a sequence are also true.
- **Proof by contradiction**
  Proof by contradiction relies on making an initial assumption and then proving that it cannot be true.

FACT

In the 1600s, French mathematician Pierre de Fermat wrote a theorem in the margin of one of his books. He claimed he had proved it, but did not reveal how, and it subsequently became known as Fermat's Last Theroem. British mathematician Andrew Wiles eventually proved the theorem by contradiction in 1994. The proof ran to more than 100 pages.

Functions

A mathematical function can be thought of as a black box with an input and an output. You put a number in, the black box does something to that number, and you are shown the output. If, for example, our input is the variable x, then the output is called $f(x)$. Functions can take many forms, including trigonometrical functions and algebraic expressions. For example, we could define the function $f(x) = x^2 + 1$ where x is a whole number. If x is 5, then $f(x) = 26$.

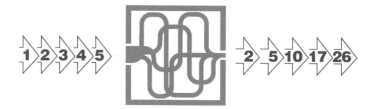

$F(x) = x^2 + 1$, so if we put x=1 into the box, we get 2 out, if we put in x=3, we get 5, and so on.

For each input, there is only one possible output. As a result, the square root operation cannot be considered a function because taking the square root of a number can give a positive or negative number.

Sets

The idea of a function is very closely tied up with the concept of sets. A set is a collection of distinct objects, which is treated as a whole. These distinct objects are called members of the set.

A set is written out inside a pair of curly brackets with the individual members separated by commas. For example:

The set of colors in the U.S. flag = {red, white, blue}

The set of the first four positive even numbers = {2, 4, 6, 8}

We can also define sets using algebraic expressions, for example:

F = {n + 1 : n is an integer; and $0 \leqslant n \leqslant 10$}

This means the set of whole numbers (integers) calculated using n + 1, when n runs from 0 to 10. We could also write this as:

F = {1, 2, 3, 4, 5, 6, 7, 8, 9, 10, 11}

KNOWING THE
ODDS

The basics of probability

What are the chances that it will rain tomorrow? One in two, perhaps, if you live in London, or one in fifteen if you live in Cairo. The area of mathematics that deals with the likelihood of certain events happening is called probability.

In mathematics, we call any situation that involves a degree of chance an experiment.

Outcomes

Each experiment has a range of possible outcomes. In the case of throwing a die, the outcomes are that the die shows 1, 2, 3, 4, 5, or 6 spots, while the outcomes of tossing a coin are heads or tails.

We refer to the complete list of possible outcomes as the sample space, and this is usually represented by a set S. For rolling a die:

$$S = \{1, 2, 3, 4, 5, 6\}$$

An event is defined as some subset of this set of all possible outcomes. For example, the event might be rolling less than a four in which case $E = \{1, 2, 3\}$.

Finally, mathematicians define the probability of an event E occurring as:

$$p(E) = \frac{\text{number of outcomes relating to event E}}{\text{total number of possible outcomes}}$$

The probability is always a fraction or a decimal between 0 and 1 (or sometimes a percentage). An event with a probability of zero will never happen, while an event with a probability of 1 is certain to happen (although this is not strictly true in the case where there are an infinite number of possible outcomes).

So if we are rolling a die, what is the probability of rolling less than a four? The answer is:

$$p(\text{rolling} < 4) = \frac{3}{6} = 0.5$$

This type of probability, based on the possible outcomes, can be distinguished from subjective probability, where an individual makes an estimate based on their own feelings and experiences of the likelihood of something happening.

Relative frequency

Imagine you decided to toss a coin 50 times and note down whether it came up heads or tails each time. The relative frequency is worked out by:

$$\text{Relative frequency} = \frac{\text{Number of times the event occurs}}{\text{Total number of times experiment carried out}}$$

If in our test, heads came up 30 times, then the relative frequency of heads is 30/50, or 0.6.

Relative frequency, which reflects a real-world experiment, may not equal the probability of something actually happening, which is about what we would expect in an idealized experiment. However, given ideal conditions and a large number of events, you would expect the relative frequency to get closer and closer to the probability.

Probability trees

A probability tree is a diagram showing the possible outcomes of an event and the probability of them occurring.

The diagram below is for an experiment involving drawing two colored balls out of a bag. There are three colored balls – red, green, and blue. The probability of drawing the first ball, where the tree starts on the left, is 0.333 for each color. Once that color has been drawn there is then a 0.5 chance of drawing either one of the other colors.

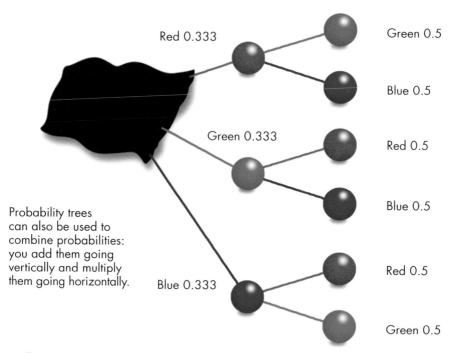

Red 0.333

Green 0.5

Blue 0.5

Green 0.333

Red 0.5

Blue 0.5

Probability trees can also be used to combine probabilities: you add them going vertically and multiply them going horizontally.

Blue 0.333

Red 0.5

Green 0.5

FACT

There is a big market for novelty bets, such as the odds of intelligent life being found elsewhere in the universe in the next 10 years. As betting shops have no way of knowing the likelihood of this happening, they simply fix the odds at a level where they can make money.

Factorial notation

In the study of probability, we often need to multiply a sequence of consecutive numbers together and so mathematicians have developed a shorthand way of writing these, called factorials. The factorial of the number n, designated by $n!$, is given by:

$$n! = n \times (n - 1) \times (n - 2) \times \ldots \times 4 \times 3 \times 2 \times 1$$

So $4! = 4 \times 3 \times 2 \times 1 = 24$ and $6! = 6 \times 5 \times 4 \times 3 \times 2 \times 1 = 720$.

Factorials come in particularly handy when evaluating something called permutations and combinations.

Permutations and combinations

Say we have four different colored balls – red, green, blue, and yellow – hidden inside a bag and we must take out two and write them down in the order we took them out. How many possible different permutations of balls are there?

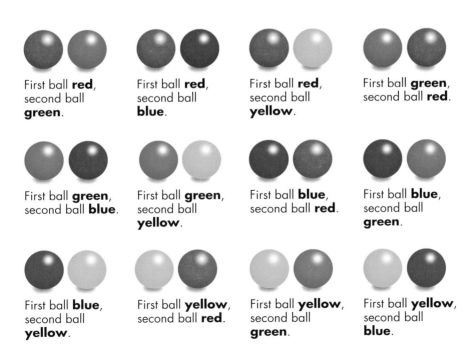

First ball **red**, second ball **green**.

First ball **red**, second ball **blue**.

First ball **red**, second ball **yellow**.

First ball **green**, second ball **red**.

First ball **green**, second ball **blue**.

First ball **green**, second ball **yellow**.

First ball **blue**, second ball **red**.

First ball **blue**, second ball **green**.

First ball **blue**, second ball **yellow**.

First ball **yellow**, second ball **red**.

First ball **yellow**, second ball **green**.

First ball **yellow**, second ball **blue**.

The answer is that there are 12 different permutations. This should be fairly obvious. When you pick out the first ball, there are four possible balls. When you pick out the second, there are only 3 remaining. Therefore the possible permutations are $4 \times 3 = 12$.

How about if the bag contained 12 different colored balls and we want to pick out 6? Of course we could write out all possible permutations in a list, but there is a simpler way.

When we pick out the first ball, there are 12 to choose from. When picking out the second, there are 11, for the third 10, for the fourth 9, for the fifth 8 and the sixth 7. This means the number of possible permutations of colors is as follows:

Number of possible permutations $= 12 \times 11 \times 10 \times 9 \times 8 \times 7$
$$= 665{,}280$$

That looks quite familiar: it's just the first part of a factorial, 12!. In fact, it is 12! without the $6 \times 5 \times 4 \times 3 \times 2 \times 1$ part. But then $6 \times 5 \times 4 \times 3 \times 2 \times 1$ is equal to 6!.

So the number of possible permutations $= 12!/6!$

In fact we can generalize the number of permutations (P) of k objects from a set of n objects as follows:

$$P(n,k) = \frac{n!}{(n-k)!}$$

Combinations

A related but subtly different idea is that of combinations. In the ball example above, we had permutations where a yellow ball was picked out first and a red second, and another where the red came out first and the yellow second. If you don't want to count repetitions, then through a similar piece of logic we can calculate the number of combinations.

Number of combinations of k objects from a set of n objects without repetition:

$$C(n,k) = \frac{n!}{k!(n-k)!}$$

Combining probabilities

Imagine you have a pack of cards and that you draw a single card from it. What is the probability that you draw out the ace of spades? From the equation earlier, we get:

$$P(\text{ace of spades}) = \frac{1}{52}$$

We now put the card back in the pile. What is the probability of now drawing out the ace of hearts? There are still 52 cards, so:

$$P(\text{ace of hearts}) = \frac{1}{52}$$

When we want to find out the probability of event A AND event B happening, we simply multiply their probailbilities:

$$P(\text{ace of spades AND ace of hearts}) = \frac{1}{52} \times \frac{1}{52} = \frac{1}{2{,}704}$$

So there is a one in 2,704 chance of pulling out both cards, i.e. it's quite unlikely.

Dependent probabilities

Let's say we want to find the probability of drawing first the ace of spades and then the ace of hearts, but we don't put the first card back in between.

We still multiply the two probabilities together, but the second probability has subtly changed because now there are only 51 cards remaining, i.e. the second event is dependent on the first.

$$P(\text{ace of spades AND ace of hearts}) = \frac{1}{52} \times \frac{1}{51} = \frac{1}{2{,}652}$$

It's still unlikely but slightly more likely than before.

The combination of dependent probabilities is written like this:

$$P \text{ (A and B)} = P(B|A)\, P(A)$$

where $P(B|A)$ means the probability that event B will happen given that event A has already happened.

What's the probability of A or B happening?

There is another instance where we might want to combine probabilities and that is in working out the probability of event A *or* event B happening. Rather than multiplying the probabilities together, we add them:

$$P \text{ (A + B)} = P(A) + P(B)$$

For example, the probability of picking the ace of spades from a fresh pack is 1 in 52; the probability of you not picking it is 51 in 52.

Add those together and you get 1, which means that it is certain that you will either pick or not pick the ace of spades.

FACT
The interest in probability came about from an interest in gambling. Seventeenth-century French mathematicians Blaise Pascal and Pierre de Fermat (of Last Theorem fame) corresponded about a dice game involving gambling on the likelihood of throwing a double six if you threw two dice 24 times in a row. Other mathematicians quickly became interested in the fledgling topic.

The birthday problem

How many people do you need in a room before the probability of two of them sharing a birthday is greater than 50%?

The answer is 23. The reasoning goes like this.

First, let's work out the probability that all their birthdays are different. The probability that the first person has a unique birthday is 1 (since we have not chosen anyone else yet). The probability that the second person has a different birthday than the first is $^{364}/_{365}$ or $1 - ^1/_{365}$. Similarly, the probability that the third person has a different birthday than the first two is $^{363}/_{365}$ or $1 - ^2/_{365}$ and so on. For the *n*th person, the probability is $1 - ^{n-1}/_{365}$.

The probability that the birthdays are all different is therefore given by all of these probabilities multiplied together, i.e.

$$p(n \text{ people, birthdays different}) = 1 \times (1-^1/_{365}) \times (1-^2/_{365}) \times (1-^3/_{365}) \dots (1 - ^{n-1}/_{365})$$

which we can rewrite as:

$$\frac{365}{365} \times \frac{364}{365} \times \frac{363}{365} \times \frac{362}{365} \times \dots \frac{(365 - n + 1)}{365}$$

This can be rewritten as:

$$\frac{365!}{365n\ (365-n)!}$$

Since we know that the probability that they do share a birthday, plus the probability that they don't has to equal 1 (i.e. it is certain that they do or don't), then we can say:

$$p(n \text{ people, birthdays same}) = 1 - \frac{365!}{365n\ (365-n)!}$$

By putting in values of *n*, we soon find that this probability exceeds 50% when *n* = 23.

Probability distributions

These are graphs that show the probability associated with every possible number of outcomes for different variables.

Discrete random variable

Imagine a statistical experiment examining how many coins each child in a school has in their pockets. The number of coins is a random variable, which changes every time a new measurement is taken. Since the number of coins can only be a whole, it is called a discrete random variable.

Continuous random variable

Other measurements, such as height, which can take any value, are called continuous random variables.

Some common probability distributions

Type of distribution	Typical diagram	Typical example
Poisson		The shape of the Poisson distribution depends on the variable λ, which is the expected number of occurrences of an event in a given interval. When λ is large, the Poisson distribution starts to look more like a normal distribution.
Normal (Gaussian)		An experiment where a continuous random variable clusters around a mean value, e.g. people's heights.
Binomial		The shape depends on the sample size and the probability. It is a discrete (not continuous) function so has a stepped appearance.

Probability and sports and games

Probability comes up often in everyday sports and games. For example, you can use your knowledge of probability to improve your chances of winning card and casino games.

Playing craps

Take the dice game of craps. The rules are as follows: the person playing (the shooter) places a bet in one of two places, called the Pass line and the Don't Pass line, on the craps table, and rolls two dice. Shooters play against the casino (the house).

The first roll of the dice is called the come-out roll, and if the shooter scores 7 or 11, the game is over. Any bets placed on the Pass line win (the shooter gets twice the bet back), while those on the Don't Pass line lose.

A come-out roll of 2, 3, or 12 is called "craps" and the game is also over. This time, bets on the Pass line lose and those on the Don't Pass line win.

For any other come-out roll (4, 5, 6, 8, 9, or 10), this figure becomes known as the point. The shooter then continues rolling. If he rolls the same figure again, the game is over and bets on the Pass line win and bets on the Don't Pass line lose. If the shooter rolls a 7, the game is over and bets on the Pass line lose, while those on the Don't Pass line win. For any other roll, the game continues.

Examining the odds

The table on the following page shows the possible outcomes from a craps game. The top row and the left column show what is showing on the dice when you make the come-out roll. The figures in the table are the values on the two dice added together.

Dice thrown	1	2	3	4	5	6
1	2	3	4	5	6	7
2	3	4	5	6	7	8
3	4	5	6	7	8	9
4	5	6	7	8	9	10
5	6	7	8	9	10	11
6	7	8	9	10	11	12

The table shows there are 36 possible outcomes from throwing a pair of dice. It also shows that there are 6 ways of throwing 7. Therefore:

$$p \text{ (throwing 7)} = 6/36 = 1/6 = 16.66\%$$

The table also shows there are two ways of throwing 11:

$$p \text{ (throwing 11)} = 2/36 = 1/18 = 5.55\%$$

From our earlier work on combining probabilities, we can say that:

$$p \text{ (throwing 7 OR throwing 11)}$$
$$= p \text{ (throwing 7)} + p \text{ (throwing 11)} = 22.22\%$$

So the chances of winning on the come-out roll alone if you have a bet on the Pass line is 22.22%.

However, you can also win by throwing 4, 5, 6, 8, 9, or 10 on the come-out roll, then the same number again before you throw a 7.

Using similar reasoning, we can work out the probabilities of throwing the other type of winning rolls (i.e. two lots of 4, 5, 6, 8, 9, or 10) as 27.07%.

Therefore the probability of rolling a winning hand in craps is 22.22% + 27.07% or 49.29%.

Adding these up along with the probability of throwing a 7 or 11 on the come-out roll gives us the total probability of winning a bet placed on the Pass line.

$$p \text{ (win on pass line)} =$$
$$22.22 + 2.77 + 4.44 + 6.31 + 6.31 + 4.44 + 2.77$$
$$= 49.29\%$$

The fact that this probability is less than 50% should make it clear to you that over the long term, the casino always wins.

Blackjack

Blackjack is another casino game that is useful for studying probability. The chances of going bust if you hit (take another card) are shown in the table below.

Existing hand	Probability of busting	Existing hand	Probability of busting
21	100%	15	58%
20	92%	14	56%
19	85%	13	39%
18	77%	12	31%
17	69%	11 or less	0%
16	62%		

The casino gets its small advantage over the players in blackjack by being the last to decide whether to take another card: the players who bust lose their bets immediately.

FACT
The probability of being dealt a blackjack in the first hand of a game is around 4.8%. The chance of getting a hand that you could stick on (18 and above) is around 27.7%.

Bayes's theorem

You are a doctor and have just started using a new test to detect for a rare disease. The test seems pretty accurate. The test corrrectly diagnoses 99% of patients who have the disease. It also diagnoses 99% of patients who don't have the disease correctly. The disease affects only 0.1% of the population. What is the probability that, following a positive test, a patient has the disease?

To find the answer we need to use Bayes's theorem, named after its discoverer, Reverend Thomas Bayes. It looks like this:

$$P(A \mid B) = \frac{P(B \mid A) \, P(A)}{P(B)}$$

where $P(A)$ is known as the marginal probability of some event A; $P(A \mid B)$ is known as the conditional probability of A, given B; $P(B \mid A)$ is the conditional probability of B given A; and $P(B)$ is the marginal probability of B.

Here, $P(A)$ is the probability of someone having the disease, in this case 0.1%. $P(B \mid A)$ is the probability that the test is positive, given that the patient has the disease, in our case 99%. $P(B)$ is the probability of getting a positive test result, which we have to calculate. It is equal to the probability that a true positive result will appear (= 99% x 0.1% = 0.099%), plus the probability that a false positive will appear (= 1% x 99.9% = 0.999%) = 1.098%.

Following a positive test, the probability that a person actually has the disease is

$$P(A \mid B) = \frac{99\% \times 0.1\%}{1.098\%} = 10.87\%$$

The test does not look so accurate now, since this means it is more likely a patient won't have the disease after a positive result.

INFINITY AND
BEYOND

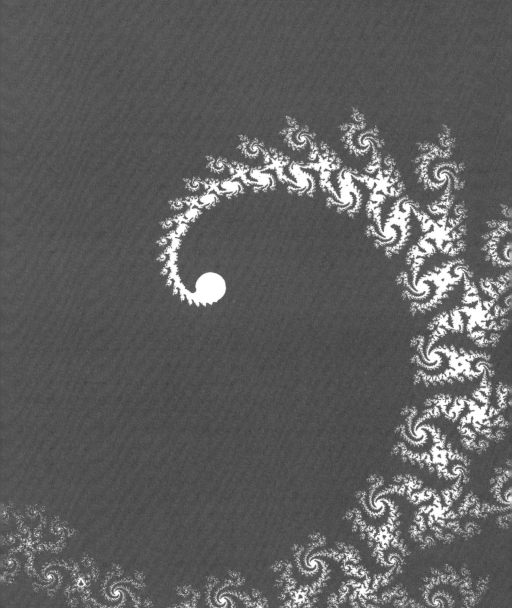

Applied mathematics

Applied math is the study of the application of mathematics to other disciplines, such as science and engineering.

For many, the heart of applied mathematics is classical mechanics, the study of the movement of objects, such as bouncing balls, the trajectories of cannonballs, and the orbits of the planets (if we discount the ideas of one Albert Einstein).

Some equations of classical mechanics

Concept	Equation
Velocity	$\mathbf{v} = \dfrac{d\mathbf{r}}{dt}$
Acceleration	$\mathbf{a} = \dfrac{d\mathbf{v}}{dt}$
Newton's second law of motion	$\mathbf{F} = m\mathbf{a} = \dfrac{d(m\mathbf{v})}{dt}$
Change in velocity under uniform acceleration	$\mathbf{v}^2 = \mathbf{u}^2 + 2as$ ($\mathbf{v}$ = final velocity, $\mathbf{u}$ = initial velocity, a = acceleration, s = distance)
Distance traveled under constant acceleration	$s = \mathbf{u}t + \frac{1}{2}at^2$ (s = distance, $\mathbf{u}$ = initial velocity, t = time, a = acceleration)

Note: Symbols in bold, italic type are vectors (see page 66).

Practical uses of applied mathematics

Applied math crops up throughout the real world. Airlines use mathematical models to decide how to construct their schedules to make best use of their planes. Investment houses use mathematical models to see how they expect the price of stocks and other investments to move over the course of time. Game theory (see "The Prisoner's Dilemma" on the opposite page) is another area where applied math is relevant to a business situation.

The Prisoner's Dilemma

In the late 1920s, the Hungarian mathematician John von Neumann published a paper called "On the Theory of Parlor Games." This looked at how mathematics could be used to analyze the outcome of games, such as poker and chess. Von Neumann was quick to spot the potential for the use of game theory in any strategic situation involving a limited number of players.

One of the best-known applications of game theory is called the Prisoner's Dilemma, as follows.

The police have two prisoners in custody, who are kept separate. They do not have enough evidence to convict either, so they offer to make deals with the two prisoners.

If one prisoner gives evidence against the other, then he may go free and the other will get a long sentence.

If both remain silent, then the prisoners will each serve a short sentence for a minor charge.

If both give evidence against the other, then they both serve a term somewhere between the two.

The prisoner's dilemma is to work out the best strategy. You can see from the table below that logically, both A and B should betray each other, since that way they will both be certain of a shorter sentence, no matter what the other prisoner decides to do. In fact, in real-world tests, more prisoners stay silent than give evidence.

Game theory is used in many areas, including the design of commercial auctions and in business strategy.

Possible outcomes		
	B is silent	**B betrays A**
A is silent	Each gets 9 months	A gets 5 years B goes free
A betrays B	B gets 5 years A goes free	Each gets 3 years

Number systems

We use a number system based on the number 10, possibly because we have 10 fingers. However, this is not the only system in use.

Decimal

This is the number system we use, based on the number 10. The position of a digit in a number determines its value. The right-most digit represents the number of ones, the digit to the left of it tens, the digit to the left hundreds, and so on as follows:

Decimal system					
100,000	10,000	1,000	100	10	1
5	6	4	3	5	7

So the number 564,357 is equal to (5 x 100,000) + (6 x 10,000) + (4 x 1,000) + (3 x 100) + (5 x 10) + (7 x 1).

The decimal system has been in use for thousands of years. Egyptian hieroglyphs show evidence of a knowledge of the system, while people in the Indus Valley were using decimal fractions perhaps as early as 3000 BC.

Binary

There is another numbering system that is in widespread use today, in the world of computers. The binary system – based on the number two – was first proposed by Indian and Chinese scholars but took its first clear steps in the work of the philosopher and scientist Francis Bacon, who suggested that letters could be encoded using binary

Francis Bacon is credited with the first version of a binary code.

representations. The binary system uses just two symbols – 0 and 1 – and again follows the same principle as the decimal and sexagesimal systems. Every move to the left in a number is a multiplication by two, as shown in the following table:

Binary system						
64	32	16	8	4	2	1
1	0	1	1	1	0	1

The binary number in the second row represents $(1 \times 64) + (0 \times 32) + (1 \times 16) + (1 \times 8) + (1 \times 4) + (0 \times 2) + (1 \times 1) = 93$.

Addition in the binary system follows these rules: $0 + 0 = 0$; $1 + 0 = 0 + 1 = 1$; and $1 + 1 = 0$ (and carry 1 to the next position to the left).

Therefore 110 + 10 = 1,000

Subtraction in binary follows these rules: $0 - 0 = 0$; $1 - 0 = 1$; $1 - 1 = 0$; and $0 - 1 = 1$ (borrowing a 1 from the next position to the left).

So 110 – 11 = 11

Binary numbers

What is the answer to the binary sum 11111 + 10101? (Hint: Write the numbers one above the other as a normal sum. Add digits starting from the right. Instead of carrying a digit over to the next column when you reach 10, do so when you reach 2.)

Imaginary numbers

The Italian mathematician Rafael Bombelli came up with the idea of imaginary numbers in the 16th century to address the problem of whether it is possible to find the square root of a negative number. He introduced the imaginary number i, equal to the square root of -1, to use when taking the root of a negative number.

Here's how. We can write -9 as 9 x -1. The square root of -9 is therefore the square root of 9 multiplied by the square root of -1. The symbol i introduced by Bombelli represents the square root of -1. So we can say that:

$$\sqrt{-9} = 3i$$

The idea of imaginary numbers was originally ridiculed, since few could grasp the concept of the square root of -1, but eventually they came to be accepted by other mathematicians. They have uses in many fields, including electrical engineering calculations involving alternating current electricity.

Complex numbers

Mathematicians have extended the idea of imaginary numbers to complex numbers, which have both a real and imaginary part; for example, $5 + 6i$ is a complex number.

FACT
The Swiss mathematician Leonhard Euler discovered a relationship linking several of the most fundamental mathematical concepts. It is called Euler's identity and is written:

$$e^{i\pi} + 1 = 0$$

Who invented calculus?

Gottfried Leibniz

Many mathematicians have made important contributions to the development of calculus, including Zeno and Archimedes.

In 1634, the French mathematician Gilles Personne de Roberval showed a method of calculating the area under a curve, while shortly after, fellow Frenchman Pierre de Fermat carried out some important work on the tangent lines to curves. However, two other mathematicians are more often associated with the invention of calculus in its current form.

In the 1660s, Sir Isaac Newton revealed his work on so-called fluxions. In particular, he looked at the velocity of a particle tracing out a curve and how the coordinates of the particle changed with time, showing that position and velocity were related through the derivative of the position with respect to time. He also investigated integration, the process of turning a derivative back into its original function.

In the 1670s, the German mathematician Gottfried Leibniz used integration to calculate the area under a curve, the first time this technique had been used. Leibniz also introduced the familiar notation of d/dx and the integration ($\int$) symbol.

Debate raged over who had invented calculus in the first place, but in reality both made enormous contributions to the development of the subject.

The beginnings of calculus

The subject of calculus begins by looking at how we calculate the gradient of a curve at a particular point: chapter 4 covers measuring the gradient of a straight line (see page 81), but you need two sets of coordinates for that.

The trick is to consider secants – lines that touch a curve at two points such as those in blue in the graph below. As the secant gets shorter, the slope of the secant gets closer to the slope of the tangent shown in green.

If we calculate the slope of the tangent line at the point $x = 5$, we can see that the slope of the secant gets closer to the gradient of the tangent, the closer we get to the point on the curve.

Imagine a point on the curve, where the x coordinate – which we shall call x_2 – is $x + h$, which is very close to the point where $x = 5$.

We can calculate what the value of y is at this point, using the formula for the curve:

$$y_2 = (x + h)^2 = x^2 + 2xh + h^2$$

$$\text{Slope of the secant} = \frac{y_2 - y_1}{x_2 - x_1}$$

and knowing that $(x_1, y_1) = (x, x^2)$, then:

$$\text{Slope} = \frac{2xh + h^2}{h} = 2x + h$$

We can write this as a limit (see page 32):

$$\text{Slope tangent} = \lim_{h \to 0} (\text{slope secant}) = \lim_{h \to 0} (2x+h)$$

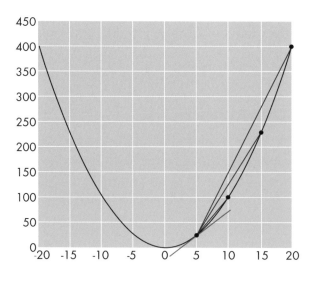

The slope of the secant gets closer to the slope of the tangent as you approach $x = 5$.

i.e. as *h* gets closer to zero, we get closer to the slope we want. But:

$$\lim_{h \to 0} (2x+h) = 2x$$

So at the point $x = 5$, the slope of the tangent is $2 \times 5 = 10$. Mathematicians write this out as:

$$f'(x) = \lim_{h \to 0} \frac{f(x+h) - f(x)}{h}$$

for any function f(*x*). Note the ' symbol in the left-hand side. This indicates that it is the first derivative of the function f(*x*), i.e. it has been derived from the original function. For a function *y*, this derivative is often written as:

$$\frac{dy}{dx}$$

Differentiation and derivatives

Finding the tangent to a curve is known as differentiation. More precisely, differentiation is the process of finding out the instantaneous rate of change of a function f(*x*) at a particular value.

In differentiation, there are some common rules for certain popular functions, as shown in the table below.

Rule/derivative	Equation
Differentiation of a polynomial term	$\frac{d(x^n)}{dx} = nx^{(n-1)}$ e.g. $\frac{d(x^3)}{dx} = 3x^2$
Linearity rule	$\frac{d(y+z)}{dx} = \frac{dy}{dx} + \frac{dz}{dx}$
Product rule	$\frac{d(yz)}{dx} = y\frac{dz}{dx} + z\frac{dy}{dx}$
Chain rule	$\frac{dy}{dz} = \frac{dy}{dx}\frac{dx}{dz}$
Derivative of a constant, c	$\frac{dc}{dz} = 0$

Integration

In the same way that we looked at differentiation by starting at finding the slope of a curve at a given point, we get to understand integration by considering the problem of working out the area underneath a curve. How do we work out the area S below?

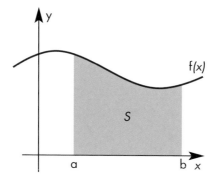

One way to do it is by dividing up the area into narrow rectangles as shown below:

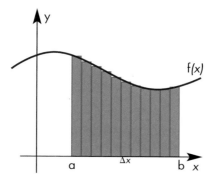

The width of the rectangles is Δx (pronounced delta x, which represents the small change in x) and the height of them is just the value of the function at that point on the curve. This last bit may sound odd, since the function clearly varies across the width of the rectangle. However, as the rectangles get smaller (i.e. we look at the limit as Δx goes to 0), then the function value gets closer and closer to the value of the function at some fixed point.

Mathematicians write this as:

$$\int_a^b f(x)dx = \lim_{n\to\infty} \sum_{i=1}^{n} f(x_1)\Delta x \quad \text{where} \quad \Delta x = \frac{b-a}{n}$$

This is called a definite integral (since it is limited to the area between *a* and *b*).

Common integrals

When you look at the table of common derivatives on page 108, you will see that differentiating a constant gives zero. This means that differentiating polynomials that are the same apart from their constant terms will give the same result:

$$\frac{d(x^3 + 3x + 2)}{dx} = 3x^2 + 3 \quad \text{and} \quad \frac{d(x^3 + 3x + 6)}{dx} = 3x^2 + 3$$

This means that when we integrate (the reverse process) then we have a problem, since $\int 3x^2 + 3 \, dx$ could be either $x^3 + 3x + 2$ or $x^3 + 3x + 6$ or indeed any other similar polynomial with a different constant.

Mathematicians get around this by adding a constant to the result of the integral, as follows:

$$\int 3x^2 + 3 \, dx = x^3 + 3x + C$$

where C is a constant.

Some common integrals

Integral	Equation
Polynomials	$\int ax^n \, dx = \frac{ax^{(n+1)}}{(n+1)} + C$
Integrals of trigonometric functions	$\int \sin x \, dx = -\cos x + C; \int \cos x \, dx = \sin x + C; \int \tan x \, dx = -\ln |\cos x| + C$
Integral of exponential function	$\int e^x \, dx = e^x + C$
Integral of reciprocal function	$\int 1/x \, dx = \ln |x| + C$

Fractals

The picture here shows an example of a fractal. One of the typical features of a fractal is that is self-similar, i.e. the whole object looks similar to one of its parts. In practice, this means that if you magnify the image and focus in on one of the purple blobs around the central figure, you will get a similarly detailed image.

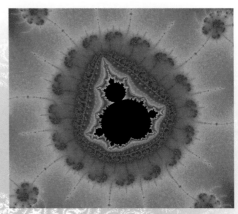

Because they appear similar at all levels of magnification, fractals are often considered to be infinitely complex.

Mandelbrot

Fractals were popularized by the French mathematician Benoît Mandelbrot in 1975 although the subject has long been a topic of discussion among mathematicians.

What is amazing is that these incredibly detailed pictures come from two deceptively simple equations:

$$z(0) = z$$

and

$$z(n+1) = z(n)^2 + z$$

I say deceptive because z is a complex rather than a real number. The colors in the image relate to the number of steps (i.e. what value of n you have reached) before the value of the second equation exceeds a certain value, say, 10. If you had the same equations but used real numbers, the pictures would not nearly be so pretty.

Fractals are relevant to many everyday phenomena. The structure of clouds and the wiggling undulations of coastlines can be modeled using fractals, for example.

Infinity

The idea of infinity, or something being without end, seems to have its origins among Indian mathematicians of the fourth century BC.

The Greeks were interested in infinity, too, but were uncomfortable with it. Aristotle, for example, decided it was real as a concept – time appeared to have no limit and you could add 1 to the highest number you could think of, for example. However, he likened infinity to the Olympic Games. He argued that you could show an outsider the athletes and the stadia but not the concept of the Games themselves, which would be impossible to grasp. He argued that infinity was equally slippery.

Countable and uncountable infinity

Have you ever wondered how many even numbers there are? Since we are talking about infinity, it should be clear that there are an infinite number of even numbers. What about odd numbers? There's an infinite number of those, too. Now, how about whole numbers (integers)? Yes, there's an infinite number of those as well.

Are all infinities the same?

Logic would say that all infinities are not the same. This brings us to the idea of countable and uncountable infinity, introduced by the German mathematician Georg Cantor.

Cantor said that an infinite set is defined as countable if it has the same cardinality as the set of all natural numbers. A set with a finite number of members is also countable. Other sets are called uncountable.

FACT
The infinity symbol was first introduced in 1657 by British mathematician John Wallis in his publication *Mathesis Universalis* (*Universal Mathematics*).

What does this mean in practice? If we can establish that there is a one-to-one relationship between the members of a particular set and the members of another infinite set, then it is countable.

Think about the natural numbers for a moment. We know that for every number n, there is another number $2n+1$, which is a member of the set of odd numbers, i.e. there is a one-to-one correspondence. Therefore the number of odd integers is the same as the number of integers. All this is very counterintuitive.

The importance of series

Hilbert's Hotel is a paradox thought up by the German mathematician David Hilbert. He envisages a hotel with an infinite number of rooms, which on a night in question is full, with an infinite number of guests.

Another prospective guest turns up and wants a room. Simple, says Hilbert, he just gets the guest in room 1 to move into room 2, the guest in room 2 to move into room 3, and so on.

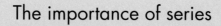

In fact, this can be extended to allow the hotel to take in a countably infinite number of new guests. You get the guest in room 1 to move into room 2, the guest in room 2 to move into room 4 and, more generally, the guest in room n to move into room $2n$. The new guests can just stay in the odd numbered rooms.

GENERAL REFERENCE

+	0.00	0.01	0.02	0.03				
1.0	.0000	.0043	.0086	.0128	.0[			
1.1	.0414	.0453	.0492	.0531	.0569			
1.2	.0792	.0828	.0864	.0899	.0934			
1.3	.1139	.1173	.1206	.1239	.1271	.13[		
1.4	.1461	.1492	.1523	.1553	.1584	.1614		
1.5	.1761	.1790	.1818	.1847	.1875	.1903		
1.6	.2041	.2068	.2095	.2122	.2148	.2175	.22[	
1.7	.2304	.2330	.2355	.2380	.2405	.2430	.2455	
1.8	.2553	.2577	.2601	.2625	.2648	.2672	.2695	
1.9	.2788	.2810	.2833	.2856	.2878	.2900	.2923	.24[
2.0	.3010	.3032	.3054	.3075	.3096	.3118	.3139	.3160
2.1	.3222	.3243	.3263	.3284	.3304	.3324	.3345	.3365
2.2	.3424	.3444	.3464	.3483	.3502	.3522	.3541	.3560
2.3	.3617	.3636	.3655	.3674	.3692	.3711	.3729	.3747
2.4	.3802	.3820	.3838	.3856	.3874	.3892	.3909	.3927
2.5	.3979	.3997	.4014	.4031	.4048	.4065	.4082	.4099
2.6	.4150	.4166	.4183	.4200	.4216	.4232	.4249	.4265
2.7	.4314	.4330	.4346	.4362	.4378	.4393	.4409	.4281
2.8	.4472	.4487	.4502	.4518	.4533	.4548	.4564	
2.9	.4624	.4639	.4654	.4669	.4683	.4698	.4712	
3.0	.4771	.4786	.4800	.4814	.4829	.4843		
3.1	.4914	.4928	.4942	.4955	.4969			
3.2	.5051	.5065	.5079	.5092				
3.3	.5185	.5198	.5211					
3.4	.5315	.5328						

General reference tables

On the following pages, you will find several tables of mathematical facts and figures that you may find helpful to refer to, including a list of common symbols, basic algebra, and a list of roots, squares, and cubes.

Common math symbols

Symbol	Meaning	Symbol	Meaning
+	Plus, addition	>	Greater than
−	Minus, subtraction	≥	Greater than or equal to
×	Times, multiplication	∝	Proportional to
÷, /	Division	∴	Therefore
=	Equal to	⇒	Implies ($a \Rightarrow b$ means that if a is true then so is b)
≈	Approximately equal to		
≠	Not equal to	∞	Infinity
√	Square root	Σ	Sum
<	Less than	∫	Integral
≤	Less than or equal to		

Powers of ten

Prefix	Power of 10	Prefix	Power of 10
exa-	10^{18}	deci-	1/10
peta-	10^{15}	centi-	1/100
tera-	10^{12}	milli-	10^{-3}
giga-	10^{9}	micro-	10^{-6}
mega-	10^{6}	nano-	10^{-9}
kilo-	1,000	pico-	10^{-12}
hecto-	100	femto-	10^{-15}
deca-	10	atto-	10^{-18}

Roots, squares, and cubes

Number	Square root	Cube root	Square	Cube
1	1	1	1	1
2	1.414	1.260	4	8
3	1.732	1.442	9	27
4	2	1.587	16	64
5	2.236	1.710	25	125
6	2.449	1.817	36	216
7	2.646	1.913	49	343
8	2.828	2	64	512
9	3	2.080	81	729
10	3.162	2.154	100	1,000
11	3.317	2.224	121	1,331
12	3.464	2.289	144	1,728
13	3.606	2.351	169	2,197
14	3.742	2.410	196	2,744
15	3.873	2.466	225	3,375
16	4	2.520	256	4,096
17	4.123	2.571	289	4,913
18	4.243	2.621	324	5,832
19	4.359	2.668	361	6,859
20	4.472	2.714	400	8,000

Basic algebra

Equation	Instruction	Result
$x + a = y + b$	Rearrange equations	$x = y + b - a$
$xa = yb$	Divide throughout by x	$a = yb/x \neq (x{\neq}0)$
$(x+a)(y+b)$	Multiply bracketed terms	$xy + xb + ay + ab$
$x^2 + xa$	Take out common factors	$x(x + a)$
$x^2 - a^2$	Difference of squares	$(x + a)(x - a)$
$1/x + 1/a$	Common denominator	$(x + a)/xa$

Symbolic language for sets

Symbol	Meaning	Example and explanation
$\in$	Is a member of	$4 \in \{1, 2, 3, 4\}$, i.e. 4 is in the set of the first four positive integers
$\notin$	Is not a member of	$2 \notin \{1, 3, 5, 7\}$, i.e. 2 is not in the set of the first four odd numbers
$\cup$	Union	This is how you add sets. $\{1,2\} \cup \{3,4\} = \{1,2,3,4\}$
$\square$	Intersection	This picks out objects that are members of both given sets, e.g. $\{1,2,3,4\} \,\square\, \{2,4,6,8\} = \{2,4\}$
$\subseteq$	Is a subset of (is contained in)	$\{1,2\} \subseteq \{1,2,3,4\}$
$\mid A \mid$	Cardinality	This gives the number of members in the set A, e.g. $\mid\{red,white,blue\}\mid = 3$
$\varnothing$	Empty set	The set with no members
$\mathbb{P}$	The set of all prime numbers	$\mathbb{P} = \{1,2,3,5,7...\}$
$\mathbb{N}$	The set of all natural numbers	$\mathbb{N} = \{1,2,3,4,5...\}$, i.e. the set of all positive integers.
$\mathbb{Z}$	The set of all integers	$\mathbb{Z} = \{..., -2, -1, 0 , 1, 2,...\}$, i.e. the set of all integers both negative and positive
$\mathbb{Q}$	The set of all rational numbers	$\mathbb{Q} = \{a/b : a, b \in \mathbb{Z}, b \neq 0\}$, i.e. the set of all numbers that can be expressed as fractions

Binary, decimal, and hexadecimal tables

Decimal	Hexa-decimal	Binary	Decimal	Hexa-decimal	Binary
000	00	00000000	026	1A	00011010
001	01	00000001	027	1B	00011011
002	02	00000010	028	1C	00011100
003	03	00000011	029	1D	00011101
004	04	00000100	030	1E	00011110
005	05	00000101	031	1F	00011111
006	06	00000110	032	20	00100000
007	07	00000111	033	21	00100001
008	08	00001000	034	22	00100010
009	09	00001001	035	23	00100011
010	0A	00001010	036	24	00100100
011	0B	00001011	037	25	00100101
012	0C	00001100	038	26	00100110
013	0D	00001101	039	27	00100111
014	0E	00001110	040	28	00101000
015	0F	00001111	041	29	00101001
016	10	00010000	042	2A	00101010
017	11	00010001	043	2B	00101011
018	12	00010010	044	2C	00101100
019	13	00010011	045	2D	00101101
020	14	00010100	046	2E	00101110
021	15	00010101	047	2F	00101111
022	16	00010110	048	30	00110000
023	17	00010111	049	31	00110001
024	18	00011000	050	32	00110010
025	19	00011001	051	33	00110011

Answers to puzzleboxes

Chapter 2

Powers of 10

7,354,267 can be written as 7.354267 x 10^6.

Reciprocal division

What is 8 divided by 0.25?
We know that 0.25 is 1/4 and that the reciprocal of 1/4 is just 4/1 or 4.
There 8 divided by 0.25 is 8 multiplied by 4 or 32.

BODMAS

What is 8 + (5 x 4^2 + 2)?
If we follow the BODMAS convention with our sum, we work out the thing
in brackets first. Then the orders, i.e. 4^2, which is 16. The multiplication next:
5 x 16 = 80. The addition next: 80 + 2 = 82. So the sum in the brackets equals
82. We then add the 8 outside the brackets to get 90.

Speedy series addition

What is a quick way of working out 1 + 2 + 3 + 4 + 5 + 6 +7 +8 + 9 + 10?
Let's write out that sum again, putting the first and the last numbers next to each
other, the second and the second last, and the third and the third last, and so on.

1 + 10 + 2 + 9 + 3 + 8 + 4 + 7 + 5 + 6 = ?

Now look at each consecutive pair of numbers. You'll notice that they each add
up to 11, so the sum we are actually looking at is actually 11 added together
five times, or 55.

Chapter 3

Trigonometric equations

We use the equation

tan A = $\dfrac{\text{opposite}}{\text{adjacent}}$

tan A = $\dfrac{5.77}{10}$ = 0.577

From the table, we can see that A=30°.
Since we know this is a right angled
triangle, one of the other angles is 90°.
As all of the angles have to add up to
180°, the remaining angle is 60°.

Trigonometrical identities

From the table in chapter 2, we have

sin 30 = 0.5, sin 45 = 0.707
cos 30 = 0.866, cos 45 = 0.707

and from our identities we can work out that

sin 15 = sin (45 − 30) = sin 45 cos 30 − cos 45 sin 30
= (0.707 x 0.866) − (0.707 x 0.5) = 0.259

cos 15 = cos (45 − 30) = cos 45 cos 30 + sin 45 sin 30
= (0.707 x 0.866) + (0.707 x 0.5) = 0.966

sin 75 = sin (45 + 30) = sin 45 cos 30 + cos 45 sin 30
= (0.707 x 0.866) + (0.707 x 0.5) = 0.966

cos 75 = cos (45 + 30) = cos 45 cos 30 − sin 45 sin 30
= (0.707 x 0.866) − (0.707 x 0.5) = 0.259

Surface area and volume of a cuboid

The volume of an individual block = 7 x 3 x 2 = 42 cm^3.
The volume of the box = 27 x 42 = 1,134 cm^3.

The box has dimensions 21 cm x 9 cm x 6 cm
(since the blocks are arranged 3 x 3 x 3).
The surface area of the box is 2 x (*lw* + *lh* + *hw*) = 2 x (21 x 9 + 21 x 6 + 9 x 6)
= 738 cm^2, so this is the minimum amount of wrapping paper you need.

Chapter 4

Expanding the brackets

$3(y − 5) = 0$ gives $3y − 15 = 0$

and

$(x − 5)(x − 7) = 0$ gives $x^2 − 5x + 35 − 7x = 0$

or

$x^2 − 12x + 35 = 0$

Equation rearranging

$5x^2 + 6x - 3 = 1/2x^2 + 3/x + 6$
Multiply throughout by x (assuming x not equal to zero) to get

$5x^3 + 6x^2 - 3x = 1/2x^3 + 3 + 6x$

Multiply both sides by two

$10x^3 + 12x^2 - 6x = x^3 + 6 + 12x$

Collect together similar terms

$9x^3 + 12x^2 - 18x - 6 = 0$

Divide both sides by three

$3x^3 + 4x^2 - 6x - 2 = 0$

Solving the puzzle

Two years ago

$F - 2 = 4 \times (B - 2) = 4B - 8$

or $F = 4B - 6$

In three years

$F + 3 = 3 \times (B + 3) = 3B + 9$

or $F = 3B + 6$

Therefore $4B - 6 = 3B + 6$, which we can rearrange to get $B = 12$. Put this into one of the equations above to get $F = 42$. Therefore the boy is 12 and the father 42.

The cookie question

You want twice as many chocolate cookies as raisin ones, so
$c = 2r$

Also we can write the cost inequality

$c \times \$0.20 + r \times \$0.15 \leqslant \$11$

Substitute the first into the second to get

$2r \times \$0.20 + r \times \$0.15 \leqslant \$11$

or rewrite to get

$r \times \$0.55 \leqslant \$11$

or $r \leqslant 20$

If r is less than or equal to 20, then 20 is the most raisin cookies you can buy. If you buy 20 raisin cookies, then you can buy a maximum of 40 chocolate cookies (since c =2r). Total number of cookies is therefore 60.

Chapter 6

Binary numbers

85 in base 10 = 1010101 in binary
255 in base 10 = 11111111 in binary

```
 11111
 10101 +
110100
```

Glossary

abacus An instrument used for calculating, which has rows of wires on which beads are slid.

calculus The branch of math that deals with the properties of derivatives and integrals of functions.

coordinate A number in a group that represents the position of a point, line, or plane.

equilateral Describing a shape whose sides are all the same length.

fractal A geometric shape whose parts have similar patterns as the whole.

function A mathematical expression that involves one or more variables.

heptagon A geometric figure that has seven straight sides and angles.

hexagon A geometric figure that has six straight sides and angles.

hyperbola A geometrical shape formed with the intersection of a cone and a plane.

hypotenuse The longest side of a right triangle.

icosahedron A solid figure that has a twenty plane faces.

integer A whole number.

octagon A geometric figure that has eight straight sides and angles.

octahedron A three-dimensional geometric shape that has eight plane faces.

parabola A symmetrical curve formed by the intersection of a cone with a plane parallel to its side.

parallelogram A geometric shape marked by a rectilinear figure with opposite sides parallel.

pentagon A geometric figure that has five straight sides and angles.

pi The approximate numerical value 3.14159, the ratio of the circumference of a circle to its diameter.

polygon A geometrical shape with at least three straight sides and angles.

polyhedron A geometrical shape with many plane faces.

polynomial The mathematical expression of more than two algebraic terms.

rhombus A geometric shape marked by opposite equal acute and obtuse angles and four equal sides.

tangent A straight line that touches a curve at a point, but does not cross it at that point.

topology The study of spatial relations that are unaffected by the continuous change of shape or size of figures.

vector A quantity that has both direction and magnitude.

For More Information

American Mathematical Society (AMS)
201 Charles Street
Providence, RI 02904-2294
(800) 321-4267
Web site: http://www.ams.org
Founded in 1888, the American Mathematical Society aims to advance the interests of mathematics around the world.

Association for Women in Mathematics (AWM)
11240 Waples Mill Road, Suite 200
Fairfax, VA 22030
(703) 934-0163
Web site: http://www.awm-math.org
The AWM is a nonprofit organization founded in 1971 whose purpose is to encourage women and girls to study and to have active careers in the mathematical sciences, and to promote equal opportunity and the equal treatment of women and girls in the mathematical sciences.

Canadian Mathematical Society (CMS)
209-1725 St. Laurent Boulevard
Ottawa, ON K1G 3V4
Canada
(613) 733-2662
Web site: http://www.cms.math.ca
The mission of the Canadian Mathematical Society is to advance important mathematical development in Canada.

Consortium for Mathematics
COMAP, Inc.
175 Middlesex Turnpike, Suite 3B
Bedford, MA 01730

(800) 772-6627

Web site: http://www.comap.com

The Consortium for Mathematics is a nonprofit organization whose mission is to improve mathematics education for students at the elementary, high school, and college level.

Mathematical Association of America (MAA)

1529 18th Street NW

Washington, DC 20036-1358

(202) 387-5200

Web site: http://www.maa.org

The Mathematical Association of America is the largest professional society that focuses on mathematics accessible at the undergraduate level.

Web Sites

Due to the changing nature of Internet links, Rosen Publishing has developed an online list of Web sites related to the subject of this book. This site is updated regularly. Please use this link to access the list:

http://www.rosenlinks.com/GCM/Math

For Further Reading

Benjamin, Arthur, Michael Shermer, Bill Nye, and Arthur Benjamin. *Secrets of Mental Math:The Mathemagician's Guide to Lightning Calculation and Amazing Math Tricks.* New York, NY:Three Rivers, 2006.

Deutsch, David. *The Beginning of Infinity: Explanations That Transform the World.* London, England: Penguin, 2012.

Gamow, George. *One, Two, Three Infinity: Facts and Speculations of Science.* New York, NY: Dover Publications, 1988.

Gray, Jeremy. *Plato's Ghost:The Modernist Transformation of Mathematics.* Princeton, NJ: Princeton University Press, 2008.

Kaku, Michio. *Physics of the Future: How Science Will Shape Human Destiny and Our Daily Lives by the Year 2100.* New York, NY: Anchor, 2012.

Marshall, Jason. *The Math Dude's Quick and Dirty Guide to Algebra.* New York, NY: St. Martin's Griffin, 2011.

Mazur, Joseph. *Zeno's Paradox: Unraveling the Ancient Mystery Behind the Science of Space and Time.* New York, NY: Penguin Group, 2007.

McKellar, Danica. *Hot X: Algebra Exposed.* New York, NY: Plume, 2010.

McKellar, Danica. *Kiss My Math: Showing Pre-algebra Who's Boss.* New York, NY: Plume, 2009.

McKellar, Danica, and Mary Lynn Blasutta. *Math Doesn't Suck: How to Survive Middle School Math Without Losing Your Mind or Breaking a Nail.* New York, NY: Plume, 2008.

Peterson, Mark A. *Galileo's Muse: Renaissance Mathematics and the Arts.* Cambridge, MA: Harvard University Press, 2011.

Plofker, Kim. *Mathematics in India.* Princeton, NJ: Princeton University Press, 2009.

Robson, Eleanor. *Mathematics in Ancient Iraq: A Social History.* Princeton, NJ: Princeton University Press, 2008.

Stewart, Ian. *Visions of Infinity:The Great Mathematical Problems.* New York, NY: Basic, 2013.

Weyl, Hermann, and Peter Pesic. *Levels of Infinity: Selected Writings on Mathematics and Philosophy.* Mineola, NY: Dover, 2013.

Index

Picture credits

The publisher would like to thank the following for permission to reproduce images:

Cover image: Shutterstock
Art Archive: pp. 12, 73; Muséum des Sciences naturelles, Belgium: p. 8; Dreamstime: p. 94; Getty Images: p. 63; iStock Photo: pp. 7, 8, 16, 25, 26, 30, 35, 38, 77, 82, 86, 91, 95, 96, 97, 99, 101, 106; John Stembridge and the American Institute of Mathematics: p. 67; NASA: p. 63; Noel Giffin and the Spanky Fractal Database: p. 114; Science and Society Picture Library: pp. 18, 109; West Semitic Research Project: p. 40.